ORACLS
A Design System for
Linear Multivariable Control

CONTROL AND SYSTEMS THEORY

A Series of Monographs and Textbooks

Editor

JERRY M. MENDEL

University of Southern California
Los Angeles, California

Associate Editors

Karl J. Astrom
Lund Institute of Technology
Lund, Sweden

Michael Athans
Massachusetts Institute of Technology
Cambridge, Massachusetts

David G. Luenberger
Stanford University
Stanford, California

Additional Volumes in Preparation

ORACLS
A Design System for
Linear Multivariable Control

ERNEST S. ARMSTRONG

National Aeronautics and Space Administration
Langley Research Center
Hampton, Virginia

and

George Washington University
NASA Graduate Engineering Program
Langley Research Center
Hampton, Virginia

MARCEL DEKKER, INC.   New York and Basel

Library of Congress Cataloging in Publication Data

Armstrong, Ernest S.
    ORACLS, a design system for linear multivariable
control.

    (Control and systems theory ; v. 10)
    Includes bibliographies.
    1. ORACLS (Computer program) I. Title.
TJ213.A66   629.8'32'0285424   80-20583
ISBN 0-8247-1239-0

MARCEL DEKKER, INC.

270 Madison Avenue, New York, New York 10016

Current printing (last digit):
10 9 8 7 6 5 4 3 2 1

PRINTED IN THE UNITED STATES OF AMERICA

To
Mary, Jeff and Laurie

PREFACE

This book describes a computational system for designing linear feedback
control laws and filters for linear time-invariant multivariable differ-
ential or difference equation state vector models. The digital FORTRAN
coded system, entitled Optimal Regulator Algorithms for the Control of
Linear Systems (ORACLS) because the majority of the design techniques
are based on optimal regulator theory, represents an application of
some of today's best numerical linear algebra procedures to implement
Linear-Quadratic-Gaussian (LQG) methodology. The collection of regulator
techniques known as LQG theory has matured in both mathematical content
and computational algorithms and is the most general time domain approach
presently available for determining multivariable control policy. The
LQG procedure provides a systematic and unified methodology for designing
multivariable control laws for systems in continuous and discrete form.

Briefly, ORACLS includes procedures for computing eigensystems of real
matrices, factored forms of matrices, the solution and least squares
approximations to solutions of certain matrix algebraic equations, the
controllability properties of a linear time-invariant system, and the
steady-state covariance matrix of an open-loop stable system forced by
white noise. Algorithms are provided for solving both the continuous and

discrete optimal linear regulator problems with noise-free measurements and the sampled-data optimal linear regulator problem. For measurement noise, duality theory and the optimal regulator algorithms are used to construct continuous and discrete Kalman-Bucy filters. Subroutines are also included which generate multivariable servomechanism control laws; that is, control laws causing the output of a linear system to follow the output of a prescribed model.

This book is a revised and updated version of an earlier ORACLS user's guide published through the NASA. Thoughts expressed in this book are those of the author and not the Federal Government. In addition to other improvements, two computational algorithms have been added to the ORACLS system for this new version--transfer matrix calculation (Chapter 4) and single-input eigenvalue placement (Chapter 5).

The ORACLS package is an outgrowth of aircraft control law research and graduate controls courses taught by the author for the George Washington University extension at Langley. Because of these applied and academic influences, it is felt that this book is equally directed towards the practicing engineer as well as use in college level courses in linear feedback control law design. Software and program listings for the complete ORACLS system can be obtained from the NASA sponsored Computer Software Management and Information Center (COSMIC), Suite 112, Barrow Hall, University of Georgia, Athens, GA 30602, by requesting program LAR-12313.

The following chapters give a detailed description of the program contents along with numerical examples to illustrate the use of ORACLS to solve selected design problems. Each ORACLS subroutine is described in a separate self-contained section in which needed mathematical symbols are defined. Therefore, only general comments need be made concerning notation. Typically, in the mathematical theory presented, lower case letters represent scalars or vectors. Upper case letters denote matrices. Any exceptions will be noted as they arise. The superscripts prime ('), plus (+), and negative unity (-1) denote matrix transpose, pseudoinverse, and inverse, respectively. The symbol ($\|\ \|$) is used for matrix norm and $\dot{x}(t)$ denotes differentiation of $x(t)$ with respect to the argument t.

The author wishes to express his thanks to Dr. E. C. Foudriat, Mr. J. R. Elliott, and Mr. W. H. Phillips, all of the Langley Research Center, for their encouragement and editorial comments while the manuscript was in preparation. Special thanks for typing services go to Mrs. Mary Edwards, Hampton, Va., and Mrs. Anda Speight, Williamsburg, Va. Finally, the author wishes to express his appreciation to the many NASA and outside users of the various versions of ORACLS for their constructive criticism and debugging aide.

ERNEST S. ARMSTRONG

CONTENTS

ONE/ THE ORACLS SYSTEM

I/ OVERVIEW OF ORACLS

The ORACLS programming system is a collection of FORTRAN-coded subroutines to formulate, manipulate, and solve various multivariable and LQG design problems [1-1]. In order to apply ORACLS, the user is required to provide an executive (driver) program which inputs the problem coefficients, formulates and selects the system subroutines to be used to solve a particular problem, and outputs desired information. ORACLS is constructed to allow the user considerable flexibility at each operational state. This flexibility is accomplished by providing primary subroutines at four levels: input-output, basic vector-matrix operations, analysis of linear time-invariant systems, and multivariable control synthesis. ORACLS provides a means of controlling program size by employing dynamic (vector) data storage. For the most part, data arrays in all ORACLS subroutines are treated as packed one-dimensional arrays which can easily be passed between subroutines without a maximum array size parameter appearing as an argument of the calling sequence. This dynamic storage capability allows program size to be specified and controlled through the user's driver program. In addition, ORACLS only loads those programs from the library which are called by the executive program, making the total machine requirements very flexible. As a result, ORACLS can be made to execute efficiently on a wide variety of computing machinery.

1

Structurally ORACLS is a modular collection of 62 subroutines with 45 primary purpose subroutines and 17 supporting. The program has approximately 6500 source statements. The ORACLS routines are written in FORTRAN IV for batch execution and have been implemented on a CDC 6000 series computer with a program length for all 62 subroutines of approximately 60K octal of 60 bit words.

The input-output category (Chapter 2) of ORACLS has subroutines for inputing (subroutine READ) and outputing (PRNT) numerical matrices. Additional subroutines allow for printing header information (RDTITL) and accumulation of output linecount information (LNCNT).

The next category (Chapter 3) has subroutines for the basic vector-matrix operations of equation (EQUATE), addition (ADD), subtraction (SUBT), and multiplication (MULT). It also contains routines for scaling (SCALE), juxtaposition (JUXTC and JUXTR), and construction of matrix norms (MAXEL and NORMS), trace (TRCE), transpose (TRANP), and null and identity matrices (NULL and UNITY).

The analysis category (Chapter 4) provides special and general purpose algorithms for computing (1) eigenvalues and eigenvectors of real matrices (EIGEN) by using the QR algorithm [1-2], (2) the relative stability of a given matrix (TESTSTA), (3) matrix factorization (FACTOR), (4) the solution of linear constant-coefficient vector-matrix algebraic equations (SYMPDS, GELIM, and SNVDEC), (5) the controllability properties of a linear time-invariant system (CTROL), (6) the steady-state covariance matrix of an open-loop stable system forced by white noise (VARANCE), (7) the transient response of continuous linear time-invariant systems (TRANSIT), and (8) the transfer matrix for a time-invariant continuous linear system (LEVIER).

2

Algorithms are provided for the solution of real matrix equations. In the equation,

$$AX = B \qquad (1\text{-}1)$$

with A positive definite, the Cholesky decomposition method [1-2] is applied (SYMPDS). Gaussian elimination (LU factorization) is used for the general case of nonsingular A (GELIM). For rectangular or singular matrices A, a singular-value decomposition procedure found in subroutine SNVDEC [1-2] can be used to factor A as

$$A = UQV' \qquad (1\text{-}2)$$

where U and V are matrices with orthonormal columns and Q is a diagonal matrix containing the ordered singular values of A. The solution of (1-1) in the least squares sense is

$$X = A^+B \qquad (1\text{-}3)$$

where A^+, the pseudo inverse of A, is given by

$$A^+ = VQ^+U' \qquad (1\text{-}4)$$

with Q^+ a diagonal matrix whose diagonal elements are the reciprocal of the nonzero singular values of A and zero otherwise. The decomposition (1-2) automatically generates an orthonormal basis for the range space of A chosen from the columns of U. A maximal rank matrix factorization for a nonnegative definite matrix can be obtained from the ORACLS subroutine FACTOR.

For solution of the matrix equation,

$$X = AXB + C \qquad (1\text{-}5)$$

the contraction mapping principle [1-3] is applied when A and B are asymptotically stable in the discrete sense (SUM). Equation (1-5) is

3

used when solving the discrete steady-state Riccati equation by Newton's method and also in the computation of the steady-state covariance matrix for a linear asymptotically stable discrete system forced by white noise [1-4].

Two subroutines are included for solving the matrix equation,

$$AX + XB = C \tag{1-6}$$

For A and B admitting a unique solution X, the method of Bartels and Stewart [1-5] is used (BARSTW). For A and B asymptotically stable in the continuous sense, a subroutine implementing the bilinear transformation approach [1-6] is also included (BILIN). Equation (1-6) is used in solving the steady-state continuous matrix Riccati equation by Newton's method, in finding the covariance statistics for continuous time-invariant systems forced by white noise, and in gain computation for observer theory [1-7].

A subroutine (CTROL) is provided which computes the controllability matrix for a linear time-invariant dynamical system and, if this matrix is found to be rank deficient, also computes the system's controllability canonical form [1-4] by application of the singular-value decomposition algorithm. Through this subroutine, the user may examine the stabilizability and detectability conditions implicit in the infinite-duration LQG methodology and, indirectly, compute minimal order state space realizations for transfer matrices.

The analysis category of ORACLS also includes subroutines (EXPSER, EXPADE, and EXPINT) for computing

$$e^{At} \quad \text{and} \quad \int_0^T e^{At}B \, dt \tag{1-7}$$

Expressions (1-7) are used in computing the transient response of linear time-invariant dynamical systems (TRANSIT). Finally, in the analysis category, Leverrier's algorithm [1-4] is employed in subroutine LEVIER to compute the transfer matrix

$$H(sI - A)^{-1}B \qquad (1-8)$$

for the linear time-invariant system

$$\dot{x}(t) = A\,x(t) + B\,u(t) \qquad (1-9)$$

with response

$$y(t) = H\,x(t) \qquad (1-10)$$

Subroutines are presented in the control law design part (Chapter 5) of ORACLS to implement some of the established techniques of time-invariant LQG methodology. For the finite duration optimal linear regulator problem using continuous dynamics (1-9) with additive Gaussian white noise, noise-free measurements (1-10), and integral performance index

$$J = E\left\{ \int_0^{t_1} \left[y(t)'\,Q\,y(t) + u(t)'\,R\,u(t) \right] dt + x'(t_1)\,P_1\,x(t_1) \right\} \qquad (1-11)$$

where E denotes expected value, the subroutine CNTNREG implements the negative exponential method [1-8] for finding both the transient and steady-state solutions to the matrix Riccati equation

$$\frac{-dP(t)}{dt} = H'QH + A'P(t) + P(t)A - P(t)BR^{-1}B'P(t) \qquad (1-12)$$

For the discrete version of this problem, the method of backward differencing is applied to find the transient and steady-state solutions to

5

the discrete Riccati equation (DISCREG). For the infinite-duration optimal linear regulator problem with noise-free measurements, a subroutine is also included to solve the steady-state Riccati equation by the Newton algorithms described by Kleinman [1-9] for continuous problems and by Hewer [1-10] for discrete problems (RICTNWT). The methods described by Armstrong [1-11] and Armstrong and Rublein [1-12] are used to compute stabilizing gains which can be used to initialize the continuous and discrete Newton iterations (CSTAB and DSTAB). A subroutine (PREFIL) is available for finding the prefilter gain to eliminate control-state cross-product terms in the quadratic performance index and another (SAMPL) computes the weighting matrices for the sampled-date optimal linear regulator problem [1-13]. For cases with measurement noise, duality theory [1-4] and the foregoing optimal regulator algorithms are used to produce solutions to the continuous and discrete Kalman-Bucy filter problems (ASYMFIL). Subroutines are included to implement the continuous [1-14] and discrete [1-15] forms of explicit (model-in-the-system) and implicit (model-in-the-performance-index) model-following theory (EXPMDFL and IMPMDFL). These subroutines generate linear control laws which cause the output of a time-invariant dynamical system to track the output of a prescribed model. Finally, an algorithm is provided for single-input eigenvalue placement (POLE).

In addition to the foregoing primary purpose subroutines, ORACLS contains supporting subroutines (Chapter 6) used predominately by the algebraic equation and eigenvalue algorithms. All subroutines of the ORACLS package are either (1) original codes by the author or (2) author-modified versions of programs contained in the VASP program [1-16], the software library of the Analysis and Computation Division at the Langley Research Center, or the numerical analysis literature. The programs coded by the

6

author and those based on the VASP program employ dynamic storage in which data arrays are treated as packed one-dimensional arrays. Other subroutines were left in their original format.

Chapter 7 presents selected design examples which illustrate the use of the foregoing ORACLS subroutines.

REFERENCES

1-1. M. Athans, "The Role and Use of the Stochastic Linear-Quadratic-Gaussian Problem in Control System Design," IEEE Trans. Autom. Control, Vol. AC-16, No. 6, Dec. 1971, pp. 529-552.

1-2. J. H. Wilkinson and C. Reinsch, Handbook for Automatic Computation, Volume II - Linear Algebra, Springer-Verlag, 1971.

1-3. L. V. Kantorovich and G. P. Akilov (D. E. Brown, trans.), Functional Analysis in Normed Spaces, (A. P. Robertson, ed.) MacMillian Co., 1964.

1-4. H. Kwakernaak and R. Sivan, Linear Optimal Control Systems, John Wiley and Sons, Inc., c.1972.

1-5. R. H. Bartels and G. W. Stewart, "Algorithm 432 - Solution of the Matrix Equation AX + XB = C," Commun. ACM, Vol. 15, No. 9, Sept. 1972, pp. 820-826.

1-6. R. A. Smith, "Matrix Equation XA + BX = C," SIAM J. Appl. Math., Vol. 16, No. 2, Mar. 1968, pp. 198-201.

1-7. D. G. Luenberger, "An Introduction to Observers," IEEE Trans. Autom. Control, Vol. AC-16, No. 6, Dec. 1971, pp. 596-602.

1-8. D. R. Vaughan, "A Negative Exponential Solution for the Matrix Riccati Equation," IEEE Trans. Autom. Control, Vol. AC-14, No. 1, Feb. 1969, pp. 72-75.

1-9. D. L. Kleinman, "On an Iterative Technique for Riccati Equation Computations," IEEE Trans. Autom. Control, Vol. AC-13, No. 1, Feb. 1968, pp. 114-115.

1-10. G. A. Hewer, "An Iterative Technique for the Computation of the Steady State Gains for the Discrete Optimal Regulator," IEEE Trans. Autom. Control, Vol. AC-16, No. 4, Aug. 1971, pp. 382-383.

1-11. E. S. Armstrong, "An Extension of Bass' Algorithm for Stabilizing Linear Continuous Constant Systems," IEEE Trans. Autom. Control, Vol. AC-20, No. 1, Feb. 1975, pp. 153-154.

1-12. E. S. Armstrong and G. T. Rublein, "A Stabilization Algorithm for Linear Discrete Constant Systems," IEEE Trans. Autom. Control, Vol. AC-21, No. 4, Aug. 1976, pp. 629-631.

1-13. E. S. Armstrong, "Series Representations for the Weighting Matrices in the Sampled-Data Optimal Linear Regulator Problem," IEEE Trans. Autom. Control, Vol. AC-23, No. 3, June 1978, pp. 478-479.

1-14. J. S. Tyler, Jr., "The Characteristics of Model-Following Systems as Synthesized by Optimal Control," IEEE Trans. Autom. Control, Vol. AC-9, No. 4, Oct. 1964, pp. 485-498.

1-15. E. S. Armstrong, "Digital Explicit Model Following With Unstable Model Dynamics," AIAA Paper No. 74-888, AIAA Mechanics and Control of Flight Conference, Aug. 1974.

1-16. J. S. White and H. Q. Lee, Users Manual for the Variable Dimension Automatic Synthesis Program (VASP), NASA TM X-2417, 1971.

TWO/ PROGRAMS FOR INPUT/OUTPUT

I/ INTRODUCTION

This chapter describes ORACLS subroutines which deal with inputing and
outputing of Hollerith and numerical data. Each of the four subroutines
described are author-modified versions of similarly titled software from
the VASP program [2-1].

II/ SUBROUTINE DESCRIPTIONS

A/ Input Hollerith Data; Define COMMON Block Data (RDTITL)

1. PURPOSE

Subroutine RDTITL reads a single card of Hollerith input which is loaded
into the array TITLE of COMMON block LINES of RDTITL and automatically
printed at the top of each page of output through the subroutine LNCNT.
The Hollerith input is typically used to define, for future reference,
the problem being solved by ORACLS.

Subroutine RDTITL also serves as a means of defining certain data blocks
important to other subroutines within ORACLS, such as information for
COMMON blocks LINES and FORM discussed in the description of LNCNT and
PRNT, respectively. The subroutines BARSTW and SNVDEC require input
parameters (EPSA, EPSB, and IAC) designating the accuracy to which solutions are to be obtained. When these programs are used internally to

other subroutines, the accuracy parameters are set to values EPSAM, EPSBM, and IACM defined by DATA statements and contained in COMMON block TOL of RDTITL. Also, convergence parameters for terminating the recursive computations in subroutines SUM, EXPSER, EXPINT, SAMPL, DISCREG, CNTNREG, and RICTNWT are internally set. Parameters SUMCV (for SUM), RICTCV (for DISCREG, CNTNREG, and RICTNWT), MAXSUM (for SUM), and SERCV (for EXPSER, EXPINT, and SAMPL) are defined by DATA statements and contained in the COMMON block CONV. The user should specify the parameters of all COMMON blocks of RDTITL on the basis of his particular computing installation and problem to be solved.

Subroutine RDTITL should be a part of every executive program provided by the user of ORACLS. If not, extraneous data appearing in the array TITLE will be printed at the top of each page of output. Also, another means of defining COMMON block data must be employed.

2. USAGE

a. Calling Sequence

CALL RDTITL

b. Input Arguments

None

c. Output Arguments

None

d. COMMON Blocks

LINES, FORM, TOL, CONV

e. Error Messages

None

f. Subroutine Employed by RDTITL

LNCNT

g. Subroutines Employing RDTITL

None

h. Concluding Remarks

None

B/ Accumulate Line Count; Page Output (LNCNT)

1. PURPOSE

Subroutine LNCNT keeps track of the number of lines printed and automatically paginates the output. Page length is controlled by the variable NLP set in the COMMON block LINES of subroutine RDTITL to 44. The variable NLP is an installation-dependent variable and may be changed as necessary. Subroutine LNCNT provides one line of print at the top of each page. This line contains 100 characters of which the first 80 are available for the programer's use and may be loaded by use of the subroutine RDTITL. The remainder contain "ORACLS PROGRAM." The 100 characters are contained in the array TITLE within RDTITL.

2. USAGE

a. Calling Sequence

CALL LNCNT(N)

b. Input Argument

N   Number of lines to be printed

c. Output Arguments

None

d. COMMON Blocks

LINES

e. Error Messages

None

f. Subroutines Employed by LNCNT

None

g. Subroutines Employing LNCNT

RDTITL, PRNT, EQUATE, TRANP, SCALE, UNITY, TRCE, ADD, SUBT, MULT, JUXTC, JUXTR, FACTOR, SUM, BILIN, BARSTW, TESTSTA, EXPSER, EXPINT, VARANCE, CTROL, TRANSIT, SAMPL, PREFIL, CSTAB, DSTAB, DISCREG, CNTNREG, RICTNWT, ASYMREG, ASYMFIL, EXPMDFL, IMPMDFL, READ1, POLE, LEVIER

h. Concluding Remarks

Subroutine LNCNT is completely internal to the ORACLS subroutines and the user need not refer to it unless he has a WRITE statement of his own. In that case, the user should put the statement CALL LNCNT(N) before each WRITE statement.

C/ Input Numerical Data (READ)

1. PURPOSE

Subroutine READ reads from one to five real matrices from cards along with their names and dimensions and prints the same information. For each matrix, a header card is first read containing a four-character title followed by two integers giving the row and column size of the matrix using format (4H,4X,2I4). Then, the matrix numerical data are read by rows using subroutine READ1 (each row of the matrix starting on a new card) using format (8F10.2). Each matrix is then automatically printed using subroutine PRNT called from READ1 and packed by columns into one-dimensional arrays with the same names.

2. USAGE

a. Calling Sequence

CALL READ(I,A,NA,B,NB,C,NC,D,ND,E,NE)

b. Input Arguments

I   Integer from 1 to 5 indicating the number of matrices to be read. For $I < 5$, the entries past I in the argument list may be omitted.

A, B, C,   Input matrices
D, E

NA, NB, NC,   Two-dimensional vectors giving the number of rows and
ND, NE   columns of the respective matrices and input through

the header card; for example,

$NA(1)$ = Number of rows of A

$NA(2)$ = Number of columns of A

If 0 is loaded for the row and column size, then the current matrix storage is unchanged, no data cards are read, and the previously stored matrix is printed.

c. Output Arguments

None

d. COMMON Blocks

None

e. Error Message

None directly from READ

However, see the description of subroutine READ1 in Chapter 6.

f. Subroutine Employed by READ

READ1

g. Subroutine Employing READ

None

h. Concluding Remarks

None

D/ Output Numerical Data (PRNT)

1. PURPOSE

Subroutine PRNT prints a single matrix with or without a descriptive heading either on the same page or on a new page. The descriptive heading, if desired, is printed before each matrix and is of the form "NAM MATRIX NA(1) ROWS NA(2) COLUMNS." The matrix is next printed by rows using format (1P7E16.7) for the first line and (3X1P7E16.7) for all subsequent lines. Format for the printing is stored in COMMON block FORM of RDTITL.

2. USAGE

a. Calling Sequence

CALL PRNT(A,NA,NAM,IOP)

b. Input Arguments

A   Matrix packed by columns in a one-dimensional array

NA   Two-dimensional vector giving the number of rows and columns of the matrix A:

           NA(1) = Number of rows of A

           NA(2) = Number of columns of A

NAM   Hollerith characters giving matrix name. Generally, NAM should contain 4 Hollerith characters and be written in the argument list as 4HXXXX. Alternatively, if 0 is inserted in the argument list for NAM, a blank name is printed.

IOP   Scalar print control parameter:

           1   Print heading and matrix on same page.

           2   Print heading and matrix after skipping to next page.

           3   Print only matrix with no heading on same page.

           4   Print only matrix with no heading after skipping to next page.

14

c. Output Arguments

None

d. COMMON Blocks

FORM, LINES

e. Error Messages

If NA(1) × NA(2) < 1 or NA(1) < 1, the message "ERROR IN PRNT MATRIX _____ HAS NA = _____ _____" is printed, and the program is returned to the calling point.

f. Subroutine Employed by PRNT

LNCNT

g. Subroutines Employing PRNT

FACTOR, SUM, BILIN, BARSTW, TESTSTA, EXPSER, EXPINT, VARANCE, CTROL, TRANSIT, SAMPL, PREFIL, CSTAB, DSTAB, DISCREG, CNTNREG, RICTNWT, ASYMREG, ASYMFIL, EXPMDFL, IMPMDFL, READ1, POLE, LEVIER

h. Concluding Remarks

None

REFERENCES

2-1. J. S. White and H. Q. Lee, Users Manual for the Variable Dimension Automatic Synthesis Program (VASP), NASA TM X-2417, 1971.

THREE/ PROGRAMS FOR VECTOR/MATRIX OPERATION

I/ INTRODUCTION

This chapter describes ORACLS subroutines which are used to implement basic vector-matrix operations. Except for the subroutines NULL, MAXEL, and NORMS, the subroutines are author-modified versions of similarly titled software from the VASP program [3-1]. Software for NORMS was obtained from the Analysis and Computation Division subprogram library at the Langley Research Center.

II/ SUBROUTINE DESCRIPTIONS

A/ Equation (EQUATE)

1. PURPOSE

Subroutine EQUATE stores a matrix A in an alternate computer location B.

2. USAGE

a. Calling Sequence

CALL EQUATE(A,NA,B,NB)

b. Input Arguments

A   Matrix packed by columns in a one-dimensional array; not
           destroyed upon return

NA   Two-dimensional vector giving the number of rows and columns

     of A:

     NA(1) = Number of rows of A

     NA(2) = Number of columns of A

     Not destroyed upon return

c. Output Arguments

B   Matrix packed by columns in a one-dimensional array. Upon

     normal return, B = A.

NB   Two-dimensional vector: NB = NA upon normal return

d. COMMON Blocks

None

e. Error Message

If NA(1) × NA(2) < 1 or NA(1) < 1, the message "DIMENSION ERROR IN

EQUATE NA = _____ _____" is printed, and the program is returned to

the calling point.

f. Subroutine Employed by EQUATE

LNCNT

g. Subroutines Employing EQUATE

FACTOR, SUM, BILIN, BARSTW, TESTSTA, EXPSER, EXPINT, VARANCE, CTROL,

TRANSIT, SAMPL, PREFIL, CSTAB, DSTAB, DISCREG, CNTNREG, RICTNWT,

ASYMREG, ASYMFIL, EXPMDFL, IMPMDFL, POLE, LEVIER

h. Concluding Remarks

None

B/ Matrix Transpose (TRANP)

1. PURPOSE

Subroutine TRANP computes the transpose A' of a given matrix A.

2. USAGE

a. Calling Sequence

CALL TRANP(A,NA,B,NB)

b. Input Arguments

A   Matrix packed by columns in a one-dimensional array; not destroyed upon return

NA   Two-dimensional vector giving the number of rows and columns of A:

           NA(1) = Number of rows of A

           NA(2) = Number of columns of A

           Not destroyed upon return

c. Output Arguments

B   Matrix packed by columns in a one-dimensional array. Upon normal return, B = A'.

NB   Two-dimensional vector: upon normal return,

           NB(1) = NA(2)

           NB(2) = NA(1)

d. COMMON Blocks

None

e. Error Message

If NA(1) × NA(2) < 1 or NA(1) < 1, the message "DIMENSION ERROR IN TRANP NA = _____ _____" is printed, and the program is returned to the calling point.

f. Subroutine Employed by TRANP

LNCNT

g. Subroutines Employing TRANP

FACTOR, SUM, BILIN, BARSTW, VARANCE, CTROL, TRANSIT, SAMPL, PREFIL, CSTAB, DSTAB, DISCREG, CNTNREG, RICTNWT, ASYMREG, ASYMFIL, EXPMDFL, IMPMDFL, POLE

h. Concluding Remarks

Subroutine TRANP is not buffered. That is, one cannot write, in general, the statement

19

```
        CALL TRANP(A,NA,A,NA)
```

to cause A to be replaced by A'.

C/ Scalar Multiplication (SCALE)

1. PURPOSE

Subroutine SCALE performs scalar multiplication on a given matrix.

2. USAGE

a. Calling Sequence

CALL SCALE(A,NA,B,NB,S)

b. Input Arguments

A   Matrix packed by columns in one-dimensional array; not

            destroyed upon return

NA   Two-dimensional vector giving the number of rows and columns

            of A:

            NA(1) = Number of rows of A

            NA(2) = Number of columns of A

            Not destroyed upon return

S   Scalar

c. Output Arguments

B   Matrix packed by columns in a one-dimensional array. Upon

            normal return, B = SA.

NB   Two-dimensional vector: NB = NA upon normal return

d. COMMON Blocks

None

e. Error Message

If NA(1) × NA(2) < 1 or NA(1) < 1, the message "DIMENSION ERROR IN

SCALE NA = _____ _____" is printed, and the program is returned to

the calling point.

f. Subroutine Employed by SCALE

LNCNT

g. Subroutines Employing SCALE

FACTOR, BILIN, EXPSER, EXPINT, VARANCE, TRANSIT, SAMPL, PREFIL, CSTAB,

DSTAB, CNTNREG, RICTNWT, ASYMREG, EXPMDFL, IMPMDFL, POLE, LEVIER

h. Concluding Remarks

None

D/ Form Identity Matrix (UNITY)

1. PURPOSE

Subroutine UNITY generates an identity matrix.

2. USAGE

a. Calling Sequence

CALL UNITY(A,NA)

b. Input Argument

NA   Two-dimensional vector giving dimension of square identity

matrix:

$NA(1) = NA(2)$ = Number of rows of A

Not destroyed upon return

c. Output Argument

A   Matrix packed by columns in a one-dimensional array. Upon

normal return, $A = I$ where I is an identity matrix of

order $NA(1)$.

d. COMMON Blocks

None

e. Error Message

If $NA(1) \neq NA(2)$, the message "DIMENSION ERROR IN UNITY NA = _____

_____" is printed, and the program is returned to the calling point.

f. Subroutine Employed by UNITY

LNCNT

g. Subroutines Employing UNITY

BILIN, EXPSER, EXPINT, TRANSIT, LEVIER

h. Concluding Remarks

None

E/ Form Null Matrix (NULL)

1. PURPOSE

Subroutine NULL generates a null matrix.

2. USAGE

a. Calling Sequence

CALL NULL(A,NA)

b. Input Argument

NA   Two-dimensional vector giving the number of rows and columns
of the desired null matrix:

$NA(1)$ = Number of rows of A

$NA(2)$ = Number of columns of A

Not destroyed upon return

c. Output Argument

A   Matrix packed by columns in a one-dimensional array. Upon
normal return, A = 0 where 0 is a null matrix of order
$NA(1) \times NA(2)$.

d. COMMON Blocks

None

e. Error Message

If $NA(1) \times NA(2) < 1$ or $NA(1) < 1$, the message "DIMENSION ERROR IN
NULL NA = _____ _____" is printed, and the program is returned to
the calling point.

22

f. Subroutine Employed by NULL

LNCNT

g. Subroutines Employing NULL

BARSTW, CNTNREG, POLE

h. Concluding Remarks

None

F/ Trace of Matrix (TRCE)

1. PURPOSE

Subroutine TRCE computes the trace of a square matrix.

2. USAGE

a. Calling Sequence

CALL TRCE(A,NA,TR)

b. Input Arguments

A   Matrix packed by columns in a one-dimensional array; not
           destroyed upon return

NA   Two-dimensional vector giving number of rows and columns
           of A:

           NA(1) = NA(2) = Order of A

           Not destroyed upon return

c. Output Argument

TR   Scalar. Upon normal return, TR = Trace of A.

d. COMMON Blocks

None

e. Error Message

If $NA(1) \neq NA(2)$, the message "TRACE REQUIRES SQUARE MATRIX NA = _____

_____" is printed, and the program is returned to the calling point.

f. Subroutine Employed by TRCE

LNCNT

23

g. Subroutines Employing TRCE

EXPSER, POLE, LEVIER

h. Concluding Remarks

None

G/ Matrix Addition (ADD)

1. PURPOSE

Subroutine ADD performs matrix addition $C = A + B$ for given

matrices   A   and   B.

2. USAGE

a. Calling Sequence

CALL ADD(A,NA,B,NB,C,NC)

b. Input Arguments

A, B   Matrices packed by columns in one-dimensional arrays; not

         destroyed upon return

NA, NB   Two-dimensional vectors giving number of rows and columns of

         respective matrices; for example,

         $NA(1)$ = Number of rows of   A

         $NA(2)$ = Number of columns of   A

         Not destroyed upon return

c. Output Arguments

C   Matrix packed by columns in a one-dimensional array. Upon

         normal return, $C = A + B$.

NC   Two-dimensional vector: upon normal return,

         $NC(1) = NA(1)$

         $NC(2) = NA(2)$

d. COMMON Blocks

None

e. Error Message

If either NA(1) ≠ NB(1), NA(2) ≠ NB(2), NA(1) < 1, or

NA(1) × NA(2) < 1, the message "DIMENSION ERROR IN ADD NA = _____

_____, NB = _____ _____" is printed, and the program is returned

to the calling point.

f. Subroutine Employed by ADD

LNCNT

g. Subroutines Employing ADD

SUM, EXPSER, EXPINT, TRANSIT, SAMPL, DSTAB, DISCREG, CNTNREG, RICTNWT,

ASYMREG, EXPMDFL, IMPMDFL, LEVIER

h. Concluding Remarks

None

H/ Matrix Subtraction (SUBT)

1. PURPOSE

Subroutine SUBT performs matrix subtraction C = A - B for given

matrices A and B.

2. USAGE

a. Calling Sequence

CALL SUBT(A,NA,B,NB,C,NC)

b. Input Arguments

A, B   Matrices packed by columns in one-dimensional arrays; not
            destroyed upon return

NA, NB   Two-dimensional vectors giving the number of rows and columns
            of respective matrices; for example,

            NA(1) = Number of rows of A

            NA(2) = Number of columns of A

            Not destroyed upon return

c. Output Arguments

C   Matrix packed by columns in a one-dimensional array. Upon normal return, C = A - B.

NC   Two-dimensional vector: upon normal return,

            NC(1) = NA(1)

            NC(2) = NA(2)

d. COMMON Blocks

None

e. Error Message

If either NA(1) $\neq$ NB(1), NA(2) $\neq$ NB(2), NA(1) < 1, or NA(1) × NA(2) < 1, the message "DIMENSION ERROR IN SUBT NA = _____ _____, NB = _____ _____" is printed, and the program is returned to the calling point.

f. Subroutine Employed by SUBT

LNCNT

g. Subroutines Employing SUBT

TRANSIT, PREFIL, CSTAB, DSTAB, DISCREG, CNTNREG, RICTNWT, ASYMREG, EXPMDFL, IMPMDFL, POLE

h. Concluding Remarks

None

I/ Matrix Multiplication (MULT)

1. PURPOSE

Subroutine MULT performs matrix multiplication C = AB for given matrices A and B.

2. USAGE

a. Calling Sequence

CALL MULT(A,NA,B,NB,C,NC)

b. Input Arguments

A, B Matrices packed by columns in one-dimensional arrays; not
    destroyed upon return

NA, NB Two-dimensional vectors giving the number of rows and columns
    of respective matrices; for example,

    $NA(1)$ = Number of rows of A

    $NA(2)$ = Number of columns of A

    Not destroyed upon return

c. Output Arguments

C   Matrix packed by columns in a one-dimensional array. Upon
    normal return, C = AB.

NC Two-dimensional vector: upon normal return,

    $NC(1) = NA(1)$

    $NC(2) = NB(2)$

d. COMMON Blocks

None

e. Error Message

If either $NA(2) \neq NB(1)$, $NA(1) < 1$, $NA(1) \times NA(2) < 1$, or
$NA(1) \times NB(2) < 1$, the message "DIMENSION ERROR IN MULT

NA = _____ _____, NB = _____ _____" is printed, and the program
is returned to the calling point.

f. Subroutine Employed by MULT

LNCNT

g. Subroutines Employing MULT

FACTOR, SUM, BILIN, EXPSER, EXPINT, VARANCE, CTROL, TRANSIT, SAMPL,
PREFIL, CSTAB, DSTAB, DISCREG, CNTNREG, RICTNWT, ASYMREG, EXPMDFL,
IMPMDFL, POLE, LEVIER

27

h. Concluding Remarks

The subroutine MULT is not buffered. That is one cannot write, in general,

CALL MULT(A,NA,B,NB,A,NA)

or

CALL MULT(A,NA,B,NB,B,NB)

to replace A or B by AB.

J/ Maximum of Matrix Elements (MAXEL)

1. PURPOSE

Subroutine MAXEL computes the maximum of the absolute values of the elements of a real matrix.

2. USAGE

a. Calling Sequence

CALL MAXEL(A,NA,ELMAX)

b. Input Arguments

A   Matrix packed by columns in a one-dimensional array; not destroyed upon return

NA   Two-dimensional vector giving the number of rows and columns of A:

NA(1) = Number of rows of A

NA(2) = Number of columns of A

Not destroyed upon return

c. Output Argument

ELMAX   Scalar. Upon normal return, ELMAX is the maximum of the absolute values of the elements of A.

d. COMMON Blocks

None

e. Error Messages

None

f. Subroutines Employed by MAXEL

None

g. Subroutines Employing MAXEL

SUM, EXPSER, EXPINT, SAMPL, DISCREG, CNTNREG, RICTNWT

h. Concluding Remarks

None

K/ Selected Matrix Norms (NORMS)

1. PURPOSE

Subroutine NORMS computes either the ℓ_1, ℓ_2 (Euclidean), or ℓ_∞ matrix norms for a real $m \times n$ matrix A stored as a variable-dimensioned two-dimensional array. The norms ℓ_1, ℓ_2, and ℓ_∞ are defined, respectively, as

$$\|A\|_1 = \max_{1 \le k \le n} \sum_{j=1}^{m} |a_{jk}| \tag{3-1}$$

$$\|A\|_2 = \left(\sum_{j=1}^{m} \sum_{k=1}^{n} a_{jk}^2 \right)^{1/2} \tag{3-2}$$

$$\|A\|_\infty = \max_{1 < j < m} \sum_{k=1}^{n} |a_{jk}| \tag{3-3}$$

2. USAGE

a. Calling Sequence

CALL NORMS(MAXROW,M,N,A,IOPT,RLNORM)

b. Input Arguments

MAXROW   Maximum first dimension of array A as given in the DIMENSION
              statement of the calling program

M   Number of rows of matrix A

N   Number of columns of the matrix A

A   Matrix whose norm is desired stored in a two-dimensional
              array

IOPT   Scalar form selector:

           1   Compute ℓ_1.

           2   Compute ℓ_2.

           3   Compute ℓ_∞.

c. Output Argument

RLNORM   Scalar. Upon normal return, RLNORM is the appropriate norm.

d. COMMON Blocks

None

e. Error Messages

None

f. Subroutines Employed by NORMS

None

g. Subroutines Employing NORMS

BILIN, EXPSER, EXPINT, SAMPL, CSTAB

h. Concluding Remarks

NORMS can be applied to matrices stored in packed one-dimensional
arrays by placing MAXROW = M in the calling sequence.

L/ Matrix Juxtaposition by Columns (JUXTC)

1. PURPOSE

Subroutine JUXTC constructs a matrix [A,B] from given matrices A
and B.

2. USAGE

a. Calling Sequence

CALL JUXTC(A,NA,B,NB,C,NC)

b. Input Arguments

A, B   Matrices packed by columns in one-dimensional arrays; not
          destroyed upon return

NA, NB   Two-dimensional vectors giving the number of rows and
          columns of the respective matrices; for example,

          NA(1) = Number of rows of A

          NA(2) = Number of columns of A

          Not destroyed upon return

c. Output Arguments

C   Matrix packed by columns in a one-dimensional array. Upon
          normal return, C = [A,B].

NC   Two-dimensional vector: upon normal return,

          NC(1) = NA(1)

          NC(2) = NA(2) + NB(2)

d. COMMON Blocks

None

e. Error Message

If either NA(1) ≠ NB(1), NA(1) < 1, NA(1) × NA(2) < 1, or

NA(2) + NB(2) < 1, the message "DIMENSION ERROR IN JUXTC

NA = _____ _____, NB = _____ _____" is printed, and the

program is returned to the calling point.

f. Subroutine Employed by JUXTC

LNCNT

g. Subroutines Employing JUXTC

TESTSTA, CTROL, DSTAB, ASYMREG, POLE

h. Concluding Remarks

None

M/ Matrix Juxtaposition by Rows (JUXTR)

1. PURPOSE

Subroutine JUXTR constructs a matrix $\begin{bmatrix} A \\ B \end{bmatrix}$ from given matrices A and B.

2. USAGE

a. Calling Sequence

CALL JUXTR(A,NA,B,NB,C,NC)

b. Input Arguments

A, B   Matrices packed by columns in one-dimensional arrays; not

           destroyed upon return

NA, NB   Two-dimensional vectors giving the number of rows and columns

           of the respective matrices; for example,

           $NA(1)$ = Number of rows of A

           $NA(2)$ = Number of columns of A

           Not destroyed upon return

c. Output Arguments

C   Matrix packed by columns in a one-dimensional array. Upon

           normal return, $C = \begin{bmatrix} A \\ B \end{bmatrix}$.

NC   Two-dimensional vector: upon normal return,

           $NC(1) = NA(1) + NB(1)$

           $NC(2) = NA(2)$

d. COMMON Blocks

None

e. Error Message

If either NA(2) $\neq$ NB(2), NA(1) < 1, NA(1) × NA(2) < 1, or

NA(2) < 1, the message "DIMENSION ERROR IN JUXTR NA = _____ _____,

NB = _____ _____" is printed, and the program is returned to the

calling point.

f. Subroutine Employed by JUXTR

LNCNT

g. Subroutines Employing JUXTR

BARSTW, CNTNREG

h. Concluding Remarks

None

REFERENCES

3-1. J. S. White and H. Q. Lee, Users Manual for the Variable
      Dimension Automatic Synthesis Program (VASP), NASA TM X-2417,
      1971.

FOUR/ PROGRAMS FOR ANALYSIS
OF CONSTANT LINEAR SYSTEMS

I/ INTRODUCTION

This chapter describes ORACLS subroutines which solve various algebraic equations occurring in linear system theory, find eigenvalues and
singular values of real matrices, evaluate matrix exponentials, perform
stability and controllability tests, calculate the transient response of
a linear system forced by a linear state variable feedback control law,
and compute the transfer matrix for a continuous linear system. Software for subroutines EIGEN, SYMPDS, GELIM, SNVDEC, and EXPADE were
obtained from the Analysis and Computation Division subprogram library
at the Langley Research Center.

II/ SUBROUTINE DESCRIPTIONS

A/ Factor Nonnegative Definite Matrix (FACTOR)

1. PURPOSE

Subroutine FACTOR computes a real $m \times n$ $(m \leq n)$ matrix D of rank m
such that a real $n \times n$ nonnegative definite matrix Q can be factored
as

$$Q = D'D \tag{4-1}$$

The method is first to perform a singular-value (eigenvalue since $Q = Q' \geqq 0$) decomposition of Q as

$$Q = PJP' \tag{4-2}$$

where J is a diagonal matrix of eigenvalues of Q, and then define D as

$$\sqrt{J} \; P' \tag{4-3}$$

after eliminating those rows of $\sqrt{J}$ with all zero elements. Eigenvalues of Q are considered negligible (zero) if they are less than $\|Q\| \times 10^{-(IAC)}$ using the matrix Euclidean norm.

2. USAGE

a. Calling Sequence

CALL FACTOR(Q,NQ,D,ND,IOP,IAC,DUMMY)

b. Input Arguments

Q   Nonnegative definite matrix packed by columns in a one-dimensional array; not destroyed upon return

NQ   Two-dimensional vector giving the number of rows and columns of Q:

           $NQ(1) = NQ(2) =$ Number of rows of Q

           Not destroyed upon return

IOP   Scalar print parameter:

           0   Return within printing results.

           Otherwise   Print Q, D, and $D'D$.

IAC   Scalar parameter to be used for zero test of eigenvalues of Q

DUMMY   Vector of working space for computations with dimension at least $n^2 + n$

c. Output Arguments

D   Matrix packed by columns in a one-dimensional array with dimension at least n^2. Upon normal return, $D'D = Q$ within numerical accuracy.

ND   Two-dimensional vector giving number of rows and columns of D: upon normal return,

$ND(1) = m$

$ND(2) = n$

d. COMMON Blocks

None

e. Error Messages

(1) If the singular-value decomposition subroutine SNVDEC fails to con-verge to a singular value after 30 iterations, the message "IN FACTOR, SNVDEC HAS FAILED TO CONVERGE TO THE _____ SINGULAR VALUE AFTER 30 ITERATIONS" is printed, and the program is returned to the calling point.

(2) If an eigenvalue of Q is greater than but close to the value $ZTEST = \|Q\| \times 10^{-(IAC)}$ (less than $16 \times ZTEST$), the message "IN FACTOR, THE MATRIX Q SUBMITTED TO SNVDEC IS CLOSE TO A MATRIX OF LOWER RANK USING ZTEST = _____ / IF THE ACCURACY IS REDUCED THE RANK MAY ALSO BE REDUCED / CURRENT RANK = _____ " is printed along with the computed singular values, and the computation continues.

f. Subroutines Employed by FACTOR

EQUATE, SNVDEC, LNCNT, TRANP, PRNT, MULT

g. Subroutines Employing FACTOR

None

h. Concluding Remarks

None

B/ Eigensystem Computation (EIGEN)

1. PURPOSE

Subroutine EIGEN computes all the eigenvalues and selected eigenvectors
of a real n × n matrix A stored as a variable-dimensioned two-
dimensional array. The input matrix is first balanced by exact similarity
transformations such that the norms of corresponding rows and columns
are nearly equal [4-1]. The balanced matrix is reduced to upper
Hessenberg form by stabilized elementary similarity transformations
[4-2]. All of the eigenvalues of the Hessenberg matrix are found by the
double shift QR algorithm [4-3]. The desired eigenvectors of the
Hessenberg matrix are then found by the inverse iteration method [4-4].

2. USAGE

a. Calling Sequence

CALL EIGEN(MAX,N,A,ER,EI,ISV,ILV,V,WK,IERR)

b. Input Arguments

MAX   Maximum first dimension of the array A as given in the
          DIMENSION statement of the calling program

N   Actual order of matrix (n)

A   Matrix whose eigenvalues and selected eigenvectors are desired
          stored in a real two-dimensional array. The contents of this
          array are destroyed upon return.

ISV   The number of eigenvalues, smallest in absolute value, for
          which eigenvectors are desired counting complex conjugates

ILV   The number of eigenvalues, largest in absolute value, for which
          eigenvectors are desired counting complex conjugates

WK   Vector of working space for computations of dimension:

          3n   if ISV + ILV = 0

          n(n+7)   otherwise

c. Output Arguments

ER   One-dimensional real array containing the real parts of the
            eigenvalues, dimensioned at least n in the calling program

EI   One-dimensional real array containing the imaginary parts of
            the eigenvalues, dimensioned at least n in the calling
            program

ISV   On output, ISV is the number of eigenvalues, smallest in
            absolute value, for which eigenvectors were computed counting
            complex conjugates

ILV   On output, ILV is the number of eigenvalues, largest in
            absolute value, for which eigenvectors were computed counting
            complex conjugates

V   Two-dimensional array containing the eigenvectors, normalized
            to unit length on a normal return. It suffices to have V
            dimensioned MAX as first dimension and n as second.

IERR   Integer error code:

            IERR = 0   Normal return

            IERR = -J   Vectors for the Jth eigenvalue did not converge
                        to an eigenvector. Appropriate column
                        (columns if Jth eigenvalue is complex) of V
                        is set to zero. If failure occurs more than
                        once, the index for the last such occurrence
                        is in IERR.

            IERR = J   The Jth eigenvalue has not been determined after
                        30 iterations of the QR algorithm.

d. COMMON Blocks

None

e. Error Messages

None. User should test IERR after return.

f. Subroutines Employed by EIGEN

BALANC, ELMHES, HQR, INVIT, ELMBAK, BALBAK

g. Subroutines Employing EIGEN

BILIN, TESTSTA, CSTAB, DSTAB, CNTNREG, ASYMREG, POLE

h. Concluding Remarks

Upon normal return, eigenvalues are stored in ascending magnitude with
complex conjugates stored with positive imaginary parts first. The
eigenvectors are packed and stored in V in the same order as their
eigenvalues appear in ER and EI. Only one eigenvector is computed for
complex conjugates (for conjugate with positive imaginary part). Upon
error exit −J, eigenvalues are correct and eigenvectors are correct for
all nonzero vectors. Upon error exit J, eigenvalues are correct but
unordered for indices IERR+1,IERR+2,...,N, and no eigenvectors are
computed.

EIGEN may be used for matrices packed as a one-dimensional array by
setting MAX = N in the calling sequence. If all eigenvectors are
desired, set ISV = N and ILV = 0.

C/ Solve AX = B, A Positive Definite (SYMPDS)

1. PURPOSE

Subroutine SYMPDS solves the matrix equation,

$$AX = B \qquad\qquad (4\text{-}4)$$

where A is a real symmetric $n \times n$ positive definite matrix and B
is a matrix of constant real vectors. The determinant of A may also
be evaluated. Solution is by Cholesky decomposition [4-4]: A is
factored as A = LDL', where L is a unit lower triangular matrix and D

is a positive definite diagonal matrix. An option is provided for computing only the Cholesky factorization of A without solving a complete matrix equation. Both A and B are stored as variable-dimensioned two-dimensional arrays.

2. USAGE

a. Calling Sequence

CALL SYMPDS(MAXN,N,A,NRHS,B,IOPT,IFAC,DETERM,ISCALE,P,IERR)

b. Input Arguments

MAXN   The maximum first dimension of A as given in the DIMENSION statement of the calling program

N   The number of rows of A (n)

A   Coefficient matrix stored as a variable-dimensioned two-dimensional array. Two types of input are possible: the first is the undecomposed coefficient matrix, and the second is the Cholesky decomposition A = LDL'. For A in unfactored form, the contents are destroyed upon return. If the factored form of A is input, A contains the upper triangular elements of the coefficient matrix and the elements of the unit lower triangular matrix L except the diagonal elements of L which are understood to be all unity. The reciprocals of the elements of D are input in the array P.

NRHS   The number of column vectors of the matrix B. No data are required if only factorization of A is desired, but NRHS still must appear as an argument of the calling sequence.

B   Two-dimensional array that must have first dimension MAXN and second dimension at least NRHS in the calling program. On input, B contains the elements of the constant vectors, and B is destroyed upon return. If only a factorization

of A is required, no data are necessary, but B still must
appear as an argument of the calling sequence.

IOPT   Integer for determinant evaluation:

0   Determinant is not evaluated.

1   Determinant is evaluated.

IFAC   Integer specifying whether or not the Cholesky decomposition of
the coefficient matrix is to be computed:

0   Cholesky decomposition for the matrix A is to be
computed and equation solved.

1   Cholesky decomposition form of A is input so no
decomposition is required.

2   Only Cholesky decomposition of A is required.

P   One-dimensional array dimensioned at least n in the calling
program. If the unfactored form of A is input, nothing
need be input for P. If the factored form of A is input,
P contains the reciprocals of the diagonal elements of D.

c. Output Arguments

A   Upon normal return, the array A contains the original elements of the matrix A and the elements of the unit lower
triangular matrix L except the diagonal elements of L
which are understood to be all unity.

B   Upon normal return, if a system of equations is to be solved,
each solution vector of X is stored over the corresponding
constant vector of the input array B.

DETERM, Determinant evaluation parameters:
ISCALE
$$\det (A) = \text{DETERM} \times 10^{(100 \times \text{ISCALE})}$$

P   Upon normal return, P contains the reciprocals of the
diagonal elements of D.

42

IERR   Error code:

        0   Normal return

        1   A is not symmetric positive definite.

d. COMMON Blocks

None

e. Error Messages

None. The parameter IERR should be tested after return.

f. Subroutines Employed by SYMPDS

None

g. Subroutines Employing SYMPDS

PREFIL, CNTNREG, RICTNWT, EXPMDFL

h. Concluding Remarks

SYMPDS may be used for matrices packed as a one-dimensional array by
setting MAXN = N in the calling sequence.

D/ Solve AX = B, A Nonsingular (GELIM)

1. PURPOSE

Subroutine GELIM solves the real matrix equation,

$$AX = B \qquad\qquad (4\text{-}5)$$

where A is $n \times n$ and nonsingular and B is a matrix of constant
real vectors. Solution is by Gaussian elimination or LU factorization
[4-4] in which A is factored as

$$PA = LU \qquad\qquad (4\text{-}6)$$

where L is a unit lower triangular matrix, U is an upper triangular
matrix, and P is a permutation matrix representing the row pivotal
strategy associated with the LU factorization. Both A and B are
stored as variable-dimensioned two-dimensional arrays.

43

2. USAGE

a. Calling Sequence

CALL GELIM(NMAX,N,A,NRHS,B,IPIVOT,IFAC,WK,IERR)

b. Input Arguments

NMAX   The maximum first dimension of the array A as given in the DIMENSION statement of the calling program

N   The number of rows of A (n)

A   Coefficient matrix stored as a variable-dimensioned two-dimensional array. Two types of input are possible: the first is the unfactored coefficient matrix, and the second is the triangular factorization A = LU. For A input in factored form, A = (L\U) should be used neglecting the unity elements of L, and the pivotal strategy employed should be input through the array IPIVOT. For A input in unfactored form, input data are destroyed.

NRHS   Number of column vectors of the matrix B

B   Two-dimensional array that must have first dimension NMAX and second dimension at least NRHS in the calling program. On input, B contains the elements of the constant vectors and is destroyed upon return.

IPIVOT   An integer array dimensioned at least n in the calling program. If the factored form of A is input, IPIVOT contains the pivotal strategy by the rule,

$$IPIVOT(I) = J$$

which states that row J of matrix A was used to pivot for the Ith unknown.

IFAC   Factorization parameter:

          0   Compute L, U, and pivotal strategy.

          1   Do not compute L, U, and pivotal strategy; the factorization and strategy are input.

WK   A one-dimensional array dimensioned at least n in the calling program and used as a work storage array

c. Output Arguments

A   Upon normal return, the unit lower and upper matrices are over-stored in A as A = (L\U) neglecting the unity elements L.

B   Upon normal return, each solution vector of X is stored over the corresponding constant vector of the input array B.

IPIVOT   Upon normal return, IPIVOT contains the pivotal strategy as previously explained.

IERR   Singularity test parameter:

          0   A is nonsingular.

          1   A is singular.

d. COMMON Blocks

None

e. Error Messages

None. Upon return the parameter IERR should be tested.

f. Subroutine Employed by GELIM

DETFAC

g. Subroutines Employing GELIM

BILIN, TRANSIT, DSTAB, CNTNREG, POLE

h. Concluding Remarks

If an LU factorization of A is desired without a complete equation solution, the subroutine DETFAC may be employed.

GELIM may be used for matrices packed as a one-dimensional array by setting NMAX = N in the calling sequence.

E/ Singular Value Decomposition (SNVDEC)

1. PURPOSE

Subroutine SNVDEC computes the singular-value decomposition [4-4] of a real $m \times n$ ($m \geqq n$) matrix A by performing the factorization,

$$A = UQV' \tag{4-7}$$

where U is an $m \times n$ matrix whose columns are n orthonormalized eigenvectors associated with the n largest eigenvalues of AA', V is an $n \times n$ matrix whose columns are the orthonormalized eigenvectors associated with the n eigenvalues of A'A, and

$$Q = \text{diag} (\sigma_1, \sigma_2, \ldots, \sigma_n) \tag{4-8}$$

where σ_i (i = 1,2,...,n) are the nonnegative square roots of the eigenvalues of A'A, called the singular values of A. Options are provided for the computation of the rank of A, singular values of A, an orthonormal basis for the null space of A, the pseudoinverse of A, and the least squares solution to

$$AX = B \tag{4-9}$$

Both A and B are stored as variable-dimensioned two-dimensional arrays. The computational procedure is described in [4-4], pages 135-151. Basically, Householder transformations are applied to reduce A to bidiagonal form after which a QR algorithm is used to find the singular values of the reduced matrix. Combining results gives the required construction.

46

2. USAGE

a. Calling Sequence

CALL SNVDEC(IOP,MD,ND,M,N,A,NOS,B,IAC,ZTEST,Q,V,IRANK,APLUS,IERR)

c. Input Arguments

IOP   Option code:

          1   The rank and singular values of A will be returned.

          2   The matrices U and V will be returned in addition
               to the information for IOP = 1.

          3   In addition to the information for IOP = 2, the least
               squares solution to AX = B will be returned.

          4   The pseudoinverse of A will be returned in addition to
               the information for IOP = 2.

          5   The least squares solution will be returned in addition
               to the information for IOP = 4.

MD   The maximum first dimension of the array A as given in the
           DIMENSION statement of the calling program

ND   Maximum first dimension of the array V

M   The number of rows of A (m)

N   The number of columns of A (n)

A   Matrix stored as a variable-dimensioned two-dimensional array.
           Input A is destroyed.

NOS   The number of column vectors of the matrix B

B   Two-dimensional array that must have row dimension at least NOS
           in the calling program. B contains the right sides of the
           equation to be solved for IOP = 3 or IOP = 5. B need
           not be input for other options but must appear in the
           calling sequence.

IAC   The number of decimal digits of accuracy in the elements of the
            matrix A. This parameter is used in the test to determine
            zero singular values and thereby the rank of A.

d. Output Arguments

A   On normal return, A contains the orthogonal matrix U except
            when IOP = 1.

B   On normal return, B contains the least squares solution for
            IOP = 3 or IOP = 5.

ZTEST   The zero test computed as $\|A\| \times 10^{-(IAC)}$ using the matrix
            Euclidean norm except when N = 1. When N = 1,

$$ZTEST = 10^{-(IAC)} \tag{4-10}$$

Q   A one-dimensional array of dimension at least n which upon
            return contains the singular values in descending order

V   A two-dimensional array that must have first dimension ND and
            second dimension at least n. Upon normal return, this array
            contains the orthogonal matrix V except when IOP = 1.
            The last N - IRANK columns of V form a basis for the
            null space of A.

IRANK   Rank of the matrix A determined as the number of nonzero
            singular values using ZTEST

APLUS   A two-dimensional array of first dimension ND and second
            dimension at least m. Upon normal return, this array con-
            tains the pseudoinverse of the matrix A. If IOP does not
            equal 4 or 5, this array need not be dimensioned, but a
            dummy parameter must appear in the calling sequence.

IERR   Error indicator:

            IERR = 0   A normal return

IERR = K > 0   The Kth singular value has not been found

after 30 iterations of the QR algorithm

procedure.

IERR = -1   Using the given IAC, A is close to a matrix

which is of lower rank and if the accuracy

is reduced, the rank of the matrix may also

be reduced.

d. COMMON Blocks

None

e. Error Messages

None. The user should examine IERR after return.

f. Subroutines Employed by SNVDEC

None

g. Subroutines Employing SNVDEC

FACTOR, CTROL, CSTAB, DSTAB, DISCREG

h. Concluding Remarks

SNVDEC may be applied to matrices stored as one-dimensional arrays by

setting MD = M and ND = N in the calling sequence.

The subroutine is internally restricted to N ≦ 150.

F/ Solve Discrete Liapunov Equation (SUM)

1. PURPOSE

Subroutine SUM evaluates the matrix series

$$X = \sum_{i=0}^{\infty} A^i B C^i \qquad (4-11)$$

where A and C are n × n and m × m real constant matrices,

respectively. The matrix B is real constant and n × m. The series is numerically summed by successively evaluating from $j = 0$ the partial sum sequence

$$\left.\begin{array}{l} S(j+1) = S(j) + U(j) \ S(j) \ V(j) \\[1em] U(j+1) = U^2(j) \\[1em] V(j+1) = V^2(j) \end{array}\right\} \tag{4-12}$$

with

$$S(0) = B \tag{4-13}$$

$$U(0) = A \tag{4-14}$$

$$V(0) = C \tag{4-15}$$

The symbol $S(j)$ represents the sum of the first 2^j terms of the power series. Evaluation of the sequence continues until the number of terms evaluated exceeds MAXSUM, which is specified in the COMMON block CONV of subroutine RDTITL, or until convergence is reached. Numerically, convergence of the $S(j)$ sequence is determined by testing the improvement in the element of $S(j)$ of largest magnitude (measured relatively if the magnitude is less than unity, and absolutely otherwise). The sequence is assumed to have converged when this improvement is less than the parameter SUMCV also found in the COMMON block CONV of subroutine RDTITL. A condition sufficient for the convergence of the power series, independent of B, is that each eigenvalue of A and C have complex modulus less than unity. When the series converges, it represents a solution X to

$$X = AXC + B \tag{4-16}$$

2. USAGE

a. Calling Sequence

CALL SUM(A,NA,B,NB,C,NC,IOP,SYM,DUMMY)

b. Input Arguments

A, B, C   Compatible matrices packed by columns in one-dimensional

            arrays. A, B, and C are destroyed upon return.

NA, NB, NC Two-dimensional vectors giving the number of rows and

            columns of the respective matrices; for example,

            NA(1) = Number of rows of A

            NA(2) = Number of columns of A

            Not destroyed upon return

IOP   Print parameter:

            0   Do not print results.

            Otherwise   Print input A, B, C, and the sum denoted

                        as X.

SYM   Logical variable:

            TRUE   Indicates A = C'

            FALSE   Otherwise

DUMMY   Vector of working space for computations with dimension at

            least maximum of $(n^2, m^2, 2nm)$

c. Output Argument

B   Upon normal return, the sum X is stored in B.

d. COMMON Block

CONV

e. Error Message

If the number of terms in the partial sum sequence $S(j)$ exceeds MAXSUM,

the message "IN SUM, THE SEQUENCE OF PARTIAL SUMS HAS EXCEEDED STAGE ____

WITHOUT CONVERGENCE" is printed.

f. Subroutines Employed by SUM

PRNT, MULT, MAXEL, ADD, EQUATE, TRANP, LNCNT

g. Subroutines Employing SUM

BILIN, VARANCE, RICTNWT, EXPMDFL

h. Concluding Remarks

A method for scaling the A and C matrices in order to possibly improve convergence is found in [4-5]. Also found in [4-5] is a bilinear transformation method for converting equation (4-16) into one solvable (assuming a unique solution exists) by the subroutine BARSTW of ORACLS.

G/ Solve Continuous Liapunov Equation (BILIN)

1. PURPOSE

Subroutine BILIN solves the matrix equation,

$$AX + XB = C \qquad (4\text{-}17)$$

where A and B are real constant matrices of dimension $n \times n$ and $m \times m$, respectively. The matrix C is real constant and of dimension $n \times m$. It is assumed that all eigenvalues of A and B have strictly negative real parts. The method of solution employs the bilinear transformation technique described by Smith [4-6], wherein it is established that equation (4-18) is equivalent to

$$X = UXV + W \qquad (4\text{-}18)$$

with

$$U = (\beta I_n - A)^{-1}(\beta I_n + A) \qquad (4\text{-}19)$$

$$V = (\beta I_m + B)(\beta I_m - B)^{-1} \qquad (4\text{-}20)$$

$$W = -2\beta(\beta I_n - A)^{-1}C(\beta I_m - B)^{-1} \qquad (4\text{-}21)$$

where β is a real scalar greater than zero and I_n and I_m are $n \times n$ and $m \times m$ identity matrices. The eigenvalues of U and V lie within the unit circle in the complex plane; therefore, the series

$$\sum_{i=0}^{\infty} U^i W V^i$$

converges and represents X. The subroutine SUM is used to evaluate the infinite series.

The user has the option of inputing β externally or having the program compute β internally based on the eigenvalues of A and B as follows. Let

$$\lambda_j = a_j + ib_j \qquad (j = 1,2,\ldots,n) \tag{4-22}$$

be eigenvalues of A ordered so that

$$|\lambda_1| \leq |\lambda_2| \leq \ldots \leq |\lambda_n| \tag{4-23}$$

Choosing β_0 from the equation,

$$\frac{(\beta_0 + a_1)^2 + b_1^2}{(\beta_0 - a_1)^2 + b_1^2} = \frac{(\beta_0 + a_n)^2 + b_n^2}{(\beta_0 - a_n)^2 + b_n^2} \tag{4-24}$$

gives, for $a_n \neq a_1$,

$$\beta_0^2 = \frac{a_1\left(a_n^2 + b_n^2\right) - a_n\left(a_1^2 + b_1^2\right)}{(a_n - a_1)} \tag{4-25}$$

When $a_n - a_1 = 0$ or $\beta_0^2 \leq 0$, instead set

$$\beta_0 = \frac{\displaystyle\sum_{j=1}^{n} \left(a_j^2 + b_j^2\right)^{1/2}}{n} \tag{4-26}$$

53

If the eigenvalue computation fails, set

$$\beta_0 = 2\|A\| \tag{4-27}$$

using the ℓ_1 matrix norm (see subroutine NORMS). Similarly, compute β_1 based on B and put

$$\beta = \frac{1}{2}(\beta_0 + \beta_1) \tag{4-28}$$

2. USAGE

a. Calling Sequence

CALL BILIN(A,NA,B,NB,C,NC,IOP,BETA,SYM,DUMMY)

b. Input Arguments

A, B, C   Compatible matrices packed by columns in one-dimensional arrays. C is destroyed upon return, but A and B are not. If A = B', no data need be entered for B and NB, but their symbols must still appear in the argument list.

NA, NB, NC   Two-dimensional vectors giving the number of rows and columns of the respective matrices; for example,

NA(1) = Number of rows of A

NA(2) = Number of columns of A

Not destroyed upon return

IOP   Two-dimensional parameter vector:

IOP(1) = 0   Do not print results.

Otherwise   Print A, B, C, BETA, and X.

IOP(2) = 0   Use input value BETA for β.

Otherwise   Compute β as previously described.

BETA   Scalar value β for numerical conditioning. No input is
                required if IOP(2) is nonzero.

SYM   Logical variable:

                TRUE   Indicates $A = B'$

                FALSE   Otherwise

DUMMY   Vector of working space for computations with dimension at
                least $(4p^2 + 2p)$ with $p = \max(n,m)$

c. Output Argument

C   Upon normal return, the solution X is stored in C.

d. COMMON Blocks

None

e. Error Messages

(1) If β is computed internally and the eigenvalue computation for A
    fails, the message "IN BILIN, THE _____ EIGENVALUE OF A HAS NOT
    BEEN DETERMINED AFTER 30 ITERATIONS" is printed. A similar message
    is printed if the eigenvalue computation for B fails.

(2) If the $(\beta I_n - A)^{-1}$ computation fails because of singularity, the
    message "IN BILIN, THE MATRIX (BETA)I - A IS SINGULAR, INCREASE
    BETA" is printed. A similar message is printed if the $(\beta I_m - B)^{-1}$
    computation fails.

f. Subroutines Employed by BILIN

LNCNT, PRNT, TRANP, EQUATE, EIGEN, SCALE, MULT, SUM, GELIM, NORMS

g. Subroutines Employing BILIN

VARANCE, CSTAB, RICTNWT

h. Concluding Remarks

For A and B in (4-17) not satisfying the requirement of eigenvalues
with strictly negative real parts, but still admitting a unique solution,
the subroutine BARSTW may be applied.

H/ Solve General Equation AX + XB = C (BARSTW)

1. PURPOSE

Subroutine BARSTW solves the matrix equation

$$AX + XB = C \qquad\qquad (4\text{-}29)$$

where A and B are real constant matrices of dimension n × n and
m × m, respectively. The matrix C is real constant and of dimension
n × m. It is assumed that

$$\lambda_i^A + \lambda_j^B \neq 0 \qquad (i = 1,2,\ldots,n; \quad j = 1,2,\ldots,m) \qquad (4\text{-}30)$$

where λ_i^A and λ_j^B are eigenvalues of A and B, respectively. The
matrix equation then has a unique solution X. The method of solution
is based on transforming A and B to real Schur form as described by
Bartels and Stewart [4-7].

2. USAGE

a. Calling Sequence

CALL BARSTW(A,NA,B,NB,C,NC,IOP,SYM,EPSA,EPSB,DUMMY)

b. Input Arguments

A, B, C   Compatible matrices packed by columns in one-dimensional
             arrays. C is destroyed upon return, but A and B are
             not. If A = B' and C = C', no data need be entered
             for B and NB, but their symbols should appear in the
             argument list.

NA, NB, NC   Two-dimensional vectors giving the number of rows and
             columns of the respective matrices; for example,

             NA(1) = Number of rows of A

             NA(2) = Number of columns of A

             Not destroyed upon return

IOP   Print option parameter:

                0   Do not print.

                Otherwise   Print A, B, C, and X.

SYM   Logical variable:

                TRUE   Indicates A = B' and C = C'

                FALSE   Otherwise

EPSA, EPSB   Convergence criteria for the reduction of A and B to

                Schur form. EPSA should be set slightly smaller than

                $10^{(-N)}$, where N is the number of significant digits in

                the elements of A. EPSB is similarly defined from B.

DUMMY   Vector of working space for computations with dimensions at

                least

                $2(n + 1)^2$ for SYM = TRUE

                $2(n + 1)^2 + 2(m + 1)^2$ for SYM = FALSE

c. Output Argument

C   Upon normal return, the solution X is stored in C.

d. COMMON Blocks

None

e. Error Message

If the reduction to Schur form fails, the message "IN BARSTW, EITHER THE
SUBROUTINE AXPXB OR ATXPXA WAS UNABLE TO REDUCE A OR B TO SCHUR FORM"
is printed, and the program is returned to the calling point.

f. Subroutines Employed by BARSTW

PRNT, TRANP, EQUATE, NULL, JUXTR, AXPXB, ATXPXA, LNCNT

g. Subroutines Employing BARSTW

VARANCE, CSTAB, DSTAB, RICTNWT, EXPMDFL

h. Concluding Remarks

If a set of equations is to be solved for a collection of C matrices with the same A and B or the Schur forms are desired along with X, the subroutines AXPXB and ATXPXA of Chapter 6 should be applied directly.

I/ Examine Matrix for Relative Stability (TESTSTA)

1. PURPOSE

Subroutine TESTSTA computes and tests the eigenvalues of a constant $n \times n$ real matrix A for stability relative to a parameter α in either the continuous or digital sense. In the continuous case, the matrix A is classified as stable if the real part of each eigenvalue is strictly less than α. Otherwise, A is classified as unstable relative to α. In the discrete case, the matrix A is classified as stable if the complex modulus of each eigenvalue is less than α. Otherwise, A is declared unstable relative to α.

2. USAGE

a. Calling Sequence

CALL TESTSTA(A,NA,ALPHA,DISC,STABLE,IOP,DUMMY)

b. Input Arguments

A   Square real matrix packed by columns in a one-dimensional
            array; not destroyed upon return

NA   Two-dimensional vector giving the number of rows and columns
            of A:

            NA(1) = Number of rows of A

            NA(2) = Number of columns of A

            Not destroyed upon return

ALPHA   Scalar α used for testing relative stability

DISC   Logical variable:

            TRUE   Test for stability in the discrete sense.

            FALSE   Test for stability in the continuous sense.

IOP   Print parameter:

            0   Do not print results.

            Otherwise   Print A, the eigenvalues of A, and stability

                            results.

DUMMY   Vector of working space for computations with dimension at

            least $(n^2 + 5n)$ where n is the order of A

c. Output Arguments

STABLE   Logical variable: upon normal return, STABLE = TRUE if the

            stability tests are satisfied; otherwise, STABLE = FALSE.

DUMMY   Upon a normal return, the eigenvalues of A are stored in the

            first 2n elements of DUMMY and packed by columns as a

            n × 2 matrix. For DISC = TRUE, the moduli of the eigen-

            values of A appear in the next n elements.

d. COMMON Blocks

None

e. Error Message

If the computation of the eigenvalues of A fails, the message

"IN TESTSTA, THE _____ EIGENVALUE OF A HAS NOT BEEN FOUND AFTER

30 ITERATIONS" is printed, and the program is returned to the calling

point.

f. Subroutines Employed by TESTSTA

EQUATE, EIGEN, JUXTC, PRNT

g. Subroutine Employing TESTSTA

ASYMREG

h. Concluding Remarks

None

J/ Matrix Exponential by Series Method (EXPSER)

1. PURPOSE

Subroutine EXPSER evaluates the matrix exponential e^{AT}, for a real
n × n matrix A and scalar T. Computation is based on the finite-
series algorithm described by Källström [4-8]. The matrix AT is scaled
by $1/2^k$ where k is a positive integer chosen so that the scaled
matrix has eigenvalues within the unit circle in the complex plane. The
series algorithm is applied to the scaled matrix until the series converges. Numerically, convergence is assumed to have occurred when the
improvement in the element of the finite series of largest magnitude
(measured relatively if the magnitude is less than unity, and absolutely
otherwise) is less than the parameter SERCV found in the COMMON block
CONV of subroutine RDTITL. Finally, the desired matrix exponential is
reconstructed from the exponential of the scaled matrix.

2. USAGE

a. Calling Sequence

CALL EXPSER(A,NA,EXPA,NEXPA,T,IOP,DUMMY)

b. Input Arguments

A   Square matrix packed by columns in a one-dimensional array; not
            destroyed upon return

NA   Two-dimensional vector giving the number of rows and columns
            of A:

            NA(1) = Number of rows of A

            NA(2) = Number of columns of A

            Not destroyed upon return

T   Scalar parameter

IOP   Print parameter:

             0   Do not print results.

             Otherwise   Print A, T, and e^{AT}.

DUMMY   Vector of working space for computations with dimension at

             least $2n^2$ where n is the order of A

c. Output Arguments

EXPA   Upon a normal return, a square matrix packed by columns in a

             one-dimensional array containing the matrix exponential e^{AT}

NEXPA   Two-dimensional column vector giving the number of rows and

             columns of e^{AT}:

             NEXPA(1) = NA(1)

             NEXPA(2) = NA(2)

d. COMMON Block

CONV

e. Error Messages

(1) The integer k is tested, and if found to be negative, the message

    "ERROR IN EXPSER, K IS NEGATIVE" is printed, and the program

    returned to the calling point.

(2) If k increases to 1000, the message "ERROR IN EXPINT, K = 1000" is

    printed, and the program is returned to the calling point.

f. Subroutines Employed by EXPSER

MAXEL, UNITY, TRCE, EQUATE, NORMS, SCALE, ADD, MULT, PRNT

g. Subroutines Employing EXPSER

TRANSIT, SAMPL, CNTNREG

h. Concluding Remarks

None

61

K/ Matrix Exponential by Padé Method (EXPADE)

1. PURPOSE

Subroutine EXPADE computes the matrix exponential e^A, where A is a real $n \times n$ matrix stored as a variable-dimensioned two-dimensional array. Computation is by the method of Padé approximation [4-9]. The matrix is first scaled by a power of 2 chosen so that the eigenvalues of the scaled matrix are within the unit circle in the complex plane. The exponential is computed for this scaled matrix using the approximation given by the ninth diagonal term in the Padé table for exponential approximations. The exponential for the original matrix is then reconstructed from the exponential of the scaled matrix.

2. USAGE

a. Calling Sequence

CALL EXPADE(MAX,N,A,EA,IDIG,WK,IERR)

b. Input Arguments

MAX Maximum first dimension of A as given in the DIMENSION statement of the calling program

N Order of matrix A (n)

A Matrix stored in a two-dimensional array with first dimension MAX and second at least N; not destroyed upon return

IDIG An estimate of the number of accurate digits in the largest elements of absolute value of e^A

WK Vector of working space for computations, dimensioned at least $n^2 + 8n$ where n is the order of A

c. Output Arguments

EA Matrix stored in a two-dimensional array with first dimension MAX and second at least N. Upon normal return, EA contains the matrix exponential e^A.

IERR   Error parameter:

    0   Normal return

    1   The sum of the absolute values of the elements of A
          is too large for any accuracy.

    2   The Padé denominator matrix is singular. Singularly
          should not occur with exact arithmetic.

d. COMMON Blocks

None

e. Error Message

None. The user should examine the parameter IERR upon return.

f. Subroutine Employed by EXPADE

GAUSEL

g. Subroutines Employing EXPADE

None

h. Concluding Remarks

None

L/ Matrix Exponential and integral (EXPINT)

1. PURPOSE

Subroutine EXPINT computes both the matrix exponential

$$e^{AT}$$

and the integral

$$\int_0^T e^{As} \, ds$$

for an n × n real matrix A and scalar T. Computation is based on the
finite-series algorithm described by Källström [4-9]. The matrix AT is
scaled by $1/2^k$ where k is a positive integer chosen so that the
scaled matrix has eigenvalues within the unit circle in the complex
plane. The series algorithms are applied to the scaled matrix until con-

vergence occurs. Numerically, convergence is assumed to have occurred in each series when the improvement in the element of the finite series of largest magnitude (measured relatively if the magnitude is less than unity, and absolutely otherwise) is less than the parameter SERCV found in the COMMON block CONV of subroutine RDTITL. Finally, the desired matrix exponential and the integral are reconstructed from the results using the scaled matrix.

2. USAGE

a. Calling Sequence

CALL EXPINT(A,NA,B,NB,C,NC,T,IOP,DUMMY)

b. Input Arguments

A   Square matrix packed by columns in a one-dimensional array;
            not destroyed upon return

NA   Two-dimensional vector giving the number of rows and columns
            of A:

            NA(1) = Number of rows of A

            NA(2) = Number of columns of A

            Not destroyed upon return

T   Scalar parameter

IOP   Print parameter:

            0   Do not print results.

            Otherwise   Print A, T, e^{AT}, and

$$\int_0^T e^{As}\, ds$$

DUMMY   Vector of working space for computations, dimensioned at
            least $2n^2$ where n is the order of A

c. Output Arguments

B   Upon normal return, square matrix packed by columns in a one-
dimensional array containing the matrix exponential

$$e^{AT}$$

NB   Two-dimensional column vector containing the row and column
size of e^{AT}:

NB(1) = NA(1)

NB(2) = NA(2)

C   Upon normal return, square matrix packed by columns in a one-
dimensional array containing the matrix

$$\int_0^T e^{As}\, ds$$

NC   Two-dimensional column vector giving the number of rows and
columns of

$$\int_0^T e^{As}\, ds$$

That is,

NC(1) = NA(1)

NC(2) = NA(2)

d. COMMON Block

CONV

e. Error Messages

(1) If k is found to be negative, the message "ERROR IN EXPINT, K IS
NEGATIVE" is printed, and the program is returned to the calling
point.

(2) If k increases to 1000, the message "ERROR IN EXPINT, K = 1000"
is printed, and the program is returned to the calling point.

65

f. Subroutines Employed by EXPINT

NORMS, SCALE, UNITY, ADD, EQUATE, MULT, LNCNT, PRNT, MAXEL

g. Subroutines Employing EXPINT

TRANSIT, SAMPL

h. Concluding Remarks

None

M/ Steady-State Variance (VARANCE)

1. PURPOSE

Subroutine VARANCE computes the steady-state variance matrix of the
state of the continuous or discrete linear time-invariant system:

Continuous

$$\dot{x}(t) = A\,x(t) + G\,\eta(t) \tag{4-31}$$

Discrete

$$x(i + 1) = A\,x(i) + G\,\eta(i) \tag{4-32}$$

where A is an $n \times n$ asymptotically stable matrix, G is an
$n \times m$ $(m \leqq n)$ matrix, and η is a zero-mean white-noise process with
continuous intensity or discrete variance Q. Following [4-10], the
intensity of a continuous white-noise process is defined as the coefficient matrix of the "δ-function" in the covariance formula. The steady-
state variance matrices, each denoted by W, satisfy the Liapunov
equations:

Continuous

$$AW + WA' = -GQG' \tag{4-33}$$

Discrete

$$W = AWA' + GQG' \tag{4-34}$$

The program provides the option of solving the continuous Liapunov equation by either subroutine BILIN or subroutine BARSTW. The discrete Liapunov equation is solved by using subroutine SUM. The computational parameter EPSA used by BARSTW is specified internally as EPSAM found in the COMMON block TOL of subroutine RDTITL.

2. USAGE

a. Calling Sequence

CALL VARANCE(A,NA,G,NG,Q,NQ,W,NW,IDENT,DISC,IOP,DUMMY)

b. Input Arguments

A, G, Q   Matrices packed by columns in one-dimensional arrays; not destroyed upon normal return except for Q which is replaced by GQG'. Storage for Q should be prescribed accordingly in the calling program.

NA, NG, NQ   Two-dimensional vectors giving number of rows and columns of the respective matrices; for example,

        NA(1) = Number of rows of A

        NA(2) = Number of columns of A

        Not destroyed upon return

IDENT   Logical variable:

        TRUE   If G is an identity matrix

        FALSE   Otherwise

        If IDENT = TRUE, no data are required in G and NG, but the variables must still appear as arguments of the calling sequence.

DISC   Logical variable:

        TRUE   If the discrete case is solved

        FALSE   For the continuous case

IOP   Three-dimensional option vector:

        IOP(1) = 0   Do not print results.

        Otherwise   Print A, G, GQG', Q, and W.

        IOP(2) = 0   Do not print from Liapunov equation
                                 subroutines employed.

        Otherwise   Print from these subroutines.

        IOP(3) = 0 and   Solve the Liapunov equation by sub-
         DISC = FALSE   routine BARSTW.

        IOP(3) $\neq$ 0 and   Solve the Liapunov equation by sub-
         DISC = FALSE   routine BILIN.

        IOP(3) is not required if DISC = TRUE.

DUMMY   Vector of working spaces for computations dimensioned at
        least:

        $2(n + 1)^2$ for DISC = FALSE and IOP(3) = 0

        $4n^2 + 2n$ for DISC = FALSE and IOP(3) $\neq$ 0

        $4n^2$ for DISC = TRUE

c. Output Arguments

W   Upon normal return, matrix packed by columns in a one-
        dimensional array holding the steady-state variance
        matrix

NW   Two-dimensional vector giving number of rows and columns
        of W: upon normal return,

        NW(1) = NA(1)

        NW(2) = NA(2)

d. COMMON Block

TOL

e. Error Messages

None

f. Subroutines Employed by VARANCE

PRNT, LNCNT, MULT, TRANP, BILIN, BARSTW, SUM

g. Subroutines Employing VARANCE

None

h. Concluding Remarks

In determining the storage allocation requirements for VARANCE, it was assumed that $m \leqq n$. For $m > n$, the matrix GQG' should be computed externally and input as Q with IDENT = TRUE.

N/ Controllability (CTROL)

1. PURPOSE

Subroutine CTROL evaluates the controllability matrix

$$C = \begin{bmatrix} B, & AB, & \ldots, & A^{n-1}B \end{bmatrix} \tag{4-35}$$

for a real constant (A,B) pair. The matrix A is $n \times n$ and B is $n \times r$ with $r \leqq n$. Options are provided to compute both the rank and singular values of C along with the controllability canonical form [4-10] for the (A,B) pair. For the optional computations, the singular-value decomposition algorithm found in subroutine SNVDEC is applied to factor the $r(n-r+1) \times n$ matrix

$$\tilde{C}' = \begin{bmatrix} B, & AB, & \ldots, & A^{n-r}B \end{bmatrix}' \tag{4-36}$$

as

$$\tilde{C}' = UQV' \tag{4-37}$$

or

$$\tilde{C} = VQU' \tag{4-38}$$

The number of nonzero elements of the $n \times n$ diagonal matrix Q determines the rank of C. Assuming C has rank $\ell \leqq n$, the first ℓ columns of V form an orthonormal basis for the range space of C and the next $(\ell + 1)$ to n columns form an orthonormal basis for the orthogonal complement to the range space of C. Hence, the pair (V'AV,V'B) represents the controllability canonical form for the original (A,B) pair.

2. USAGE

a. Calling Sequence

CALL CTROL(A,NA,B,NB,C,NC,IOP,IAC,IRANK,DUMMY)

b. Input Arguments

A, B   Matrices packed by columns in one-dimensional arrays; not destroyed upon return

NA, NB   Two-dimensional vectors giving the number of rows and columns of respective matrices; for example,

NA(1) = Number of rows of A

NA(2) = Number of columns of A

Not destroyed upon return

IOP   Five-dimensional option vector:

IOP(1) = 0   Do not print A, B, and C.

Otherwise   Print A, B, and C.

IOP(2) = 0   Return after computing C.

Otherwise   Compute rank of C.

IOP(3) = 0   Do not print rank of C, zero test employed to determine rank, and singular values of C.

Otherwise   Print these data.

IOP(4) = 0   Return after rank computation.

Otherwise   Compute the controllability canonical form.

IOP(5) = 0   Return without printing controllability form.

Otherwise   Print V'AV, V'B, and V' before returning.

70

IAC   Parameter used to specify zero test for rank computation.

Singular values are considered zero if they do not exceed

$ZTEST = \|\tilde{C}\| \times 10^{-(IAC)}$ using the matrix Euclidean norm.

DUMMY   Vector of working space for computations dimensioned at least:

$nr(n+1) = K$ To compute C only

$\max[K, n^2+n+nr(n-r+1)] = L$ To compute rank of C

$\max(L, 3n^2)$ To compute controllability

                                             canonical form

c. Output Arguments

C   Matrix packed by columns in a one-dimensional array dimensioned
at least n^2r. Upon normal return, C contains the controllability matrix for the (A,B) pair.

NC   Two-dimensional vector giving the number of rows and columns
of C: upon normal return,

$NC(1) = NA(1) = n$

$NC(2) = NA(1) \times NB(1) = nr$

IRANK   Upon normal return, scalar giving the rank of C

DUMMY   Upon normal return for rank computation only, the first
$nr(n-r+1)$ elements contain the matrix U, the next n elements contain the singular values of $\tilde{C}'$, and the next n^2
elements contain the matrix V. Upon normal return for the
canonical form computation, the first n^2 elements contain
the matrix V', the next nr elements contain the matrix $V'B$,
and, after the first $2n^2$ elements of DUMMY, the next n^2
elements contain the matrix $V'AV$. All matrices are packed
by columns into the corresponding sections of the one-
dimensional array DUMMY.

d. COMMON Blocks

None

e. Error Messages

(1) If SNVDEC fails to compute the singular values of $\tilde{C}$, the message "IN
CTROL, SNDVEC HAS FAILED TO CONVERGE TO THE _____ SINGULAR VALUE
AFTER 30 ITERATIONS" is printed, and the program is returned to
the calling point.

(2) If an eigenvalue of Q is greater than but close to ZTEST (see
subroutine SNVDEC), the message "IN CTROL, THE MATRIX SUBMITTED TO
SNVDEC USING ZTEST = _____ IS CLOSE TO A MATRIX WHICH IS OF LOWER
RANK / IF THE ACCURACY IS REDUCED THE RANK MAY ALSO BE REDUCED/
CURRENT RANK = _____" is printed, and the computation continues.

f. Subroutines Employed by CTROL

EQUATE, MULT, JUXTC, PRNT, LNCNT, TRANP, SNVDEC

g. Subroutines Employing CTROL

None

h. Concluding Remarks

Employing duality theory, CTROL may also be used to compute the reconstructibility canonical form [4-10]. By combining controllability and
reconstructibility results from CTROL, the full canonical structure
theorem [4-11] can be implemented.

O/ Transient Response (TRANSIT)

1. PURPOSE

Subroutine TRANSIT computes and prints the transient response of the
time-invariant continuous or discrete system.

Continuous

$$\dot{X}(t) = AX(t) + B\,U(t) \tag{4-39}$$

Discrete

$$X(i + 1) = AX(i) + B U(i) \qquad\qquad (4\text{-}40)$$

together with   Y   and   U   where

$$Y = HX + GU \qquad\qquad (4\text{-}41)$$

$$U = -FX + V \qquad\qquad (4\text{-}42)$$

for given

$$X(0) = X_0 \qquad\qquad (4\text{-}43)$$

from the initial time or stage zero to an input final time or stage.
The matrices   A,   B,   X, and   H   are dimensioned   n × n,   n × r   (r ≦ n),
n × p   (p ≦ n), and   m × n   (m ≦ n), respectively, with the other matrices
compatible.   If the matrix   (A - BF)   is asymptotically stable in the
appropriate continuous or discrete sense, the steady-state value of   X
given by

Continuous

$$X = -(A - BF)^{-1}BV \qquad\qquad (4\text{-}44)$$

Discrete

$$X = \left[I - (A - BF)\right]^{-1}BV \qquad\qquad (4\text{-}45)$$

where   I   is an   n × n   identity matrix, is computed and printed except
when   V   is a null matrix.

2. USAGE

a. Calling Sequence

CALL TRANSIT(A,NA,B,NB,H,NH,G,NG,F,NF,V,NV,T,X,NX,DISC,STABLE,IOP,DUMMY)

b. Input Arguments

A, B, H, G, F, V	Compatible matrices packed by columns in one-dimensional arrays; not destroyed upon normal return
NA, NB, NH, NG, NF, NV	Two-dimensional vectors giving the number of rows and columns of respective matrices; for example,

NA(1) = Number of rows of A

NA(2) = Number of columns of A

Not destroyed upon return

T   Two-dimensional vector:

      T(1) is the final time in the continuous case given as an integer multiple of T(2).

      T(2) is a time point within $(0, T(1)]$; printing is at multiples of T(2).

      The input T is not required if the response to a discrete system is to be computed, but the argument must still appear in the calling sequence; not destroyed upon normal return.

X   Matrix packed by columns in a one-dimensional array containing the initial value X_0 at time or stage zero.

      X is destroyed upon normal return.

NX   Two-dimensional vector giving the number of rows and columns of X:

      NX(1) = NA(1)

      NX(2) = NV(2)

      Not destroyed upon normal return

DISC   Logical variable:

      TRUE If the response to the discrete system is required

      FALSE For the response to the continuous system

STABLE Logical variable:

      TRUE   If (A - BF) is asymptotically stable in the appropriate sense

      STABLE = FALSE Otherwise

IOP   Four-dimensional option vector:

     $IOP(1) = 0$ If H is a null matrix

     $IOP(1) = 1$ If H is an identity matrix

     $IOP(1) \neq 0$ or 1 For other H matrices

     $IOP(2) = 0$ If G is a null matrix

     $IOP(2) = 1$ If G is an identity matrix

     $IOP(2) \neq 0$ or 1 For other G matrices

     $IOP(3) = 0$ If V is a null matrix

     $IOP(3) = 1$ If V is an identity matrix

     $IOP(3) \neq 0$ or 1 For other V matrices

     $IOP(4)$ is the terminal stage for the response to a discrete system; not required if the continuous system response is computed (DISC = FALSE).

DUMMY   Vector of working space for computations, dimensioned at least $7n^2$ where n is the order of A.

c. Output Arguments

X   Upon normal return, the value of X at time $T(1)$ or state $IOP(4)$

DUMMY   Upon normal return, the first np elements of DUMMY contain the steady-state value of X, packed by columns in a one-dimensional array, when STABLE = TRUE and $IOP(3) \neq 0$.

d. COMMON Blocks

None

e. Error Message

When computation of the continuous steady-state X values is attempted and $(A - BF)$ is found to be singular, the message "IN TRANSIT, THE

MATRIX A-BF SUBMITTED TO GELIM IS SINGULAR" is printed, and the program is returned to the calling point. For the discrete case, a similar message concerning the matrix $[I - (A - BF)]$ is also printed.

f. Subroutines Employed by TRANSIT

PRNT, EXPSER, EXPINT, MULT, EQUATE, LNCNT, SCALE, ADD, TRANP, GELIM, UNITY, SUBT

g. Subroutines Employing TRANSIT

None

h. Concluding Remarks

When the matrices H, G, or V take on their special null or identity matrix values, no data need be input, but they must still appear as arguments of the calling sequence.

P/ Transfer Matrix (LEVIER)

1. PURPOSE

Subroutine LEVIER evaluates the transfer matrix

$$H(sI - A)^{-1}B \qquad (4\text{-}46)$$

for the linear time-invariant system

$$\dot{x}(t) = Ax(t) + Bu(t) \qquad (4\text{-}47)$$

with output

$$y(t) = Hx(t) \qquad (4\text{-}48)$$

In $(4\text{-}46)$, s is the Laplace transform parameter and I is an $n \times n$ identity matrix corresponding to the dimension of A. The matrices B and H are dimensioned $n \times r$ $(r \leqq n)$ and $m \times n$ $(m \leqq n)$, respectively.

The application of Leverrier's $[4\text{-}10]$ algorithm gives

$$(sI - A)^{-1} = \frac{C(s)}{D(s)} \qquad (4\text{-}49)$$

76

where

$$D(s) = DET\ (sI - A) = s^n + d_1 s^{n-1} + \ldots + d_{n-1}s + d_n \quad (4\text{-}50)$$

and

$$C(s) = s^{n-1}C_0 + s^{n-2}C_1 + \ldots + sC_{n-2} + C_{n-1} \quad (4\text{-}51)$$

The $n \times n$ matrices C_i, $(i = 0,1,\ldots,n-1)$, and the scalars d_i, $(i = 1,\ldots,n)$, are computed sequentially through

$$C_0 = I$$

$$
\begin{aligned}
d_1 &= -\text{TRACE}\ (A) & C_1 &= C_0 A + d_1 I \\
d_2 &= -\text{TRACE}\ (C_1 A) & C_2 &= C_1 A + d_2 I \\
d_3 &= -\tfrac{1}{3}\ \text{TRACE}\ (C_2 A) & C_3 &= C_2 A + d_3 I \\
&\ \ \vdots & &\ \ \vdots \\
d_k &= -\tfrac{1}{k}\ \text{TRACE}\ (C_{k-1} A) & C_k &= C_{k-1} A + d_k I \\
&\ \ \vdots & &\ \ \vdots \\
d_{n-1} &= -\tfrac{1}{(n-1)}\ \text{TRACE}\ (C_{n-2} A) & C_{n-1} &= C_{n-2} A + d_{n-1} I \\
d_n &= -\tfrac{1}{n}\ \text{TRACE}\ (C_{n-1} A) & &
\end{aligned}
$$

$$(4\text{-}52)$$

Carrying the sequence one additional step gives

$$C_n = C_{n-1} A + d_n I = 0 \quad (4\text{-}53)$$

from the Cayley-Hamilton theorem [4-10]. Condition (4-53) provides a measure of the round-off error accumulated in (4-52) which can be significant.

Using (4-51), the $m \times r$ matrix (4-52) can be expressed as

$$H(sI - A)^{-1}B = \left(s^{n-1}HC_0 B + s^{n-2}HC_1 B + \ldots + \right.$$

$$\left. sHC_{n-2}B + HC_{n-1}B\right)\Big/D(s) \quad (4\text{-}54)$$

2. USAGE

a. Calling Sequence

CALL LEVIER(A,NA,B,NB,H,NH,C,NC,HCB,NHCB,D,ND,BIDENT,HIDENT,IOP,DUMMY)

b. Input Arguments

A, B, H   Matrices packed by columns in one-dimensional arrays of dimension n^2, nr, and mn, respectively, where n is the order of A, r the number of columns of B, and m the number of rows of H; not destroyed upon normal return.

NA, NB, NH   Two-dimensional vectors giving the number of rows and columns of respective matrices; for example,

NA(1) = Number of rows of A

NA(2) = Number of columns of A

BIDENT   Logical variable:

TRUE   B is an identity matrix

FALSE   Otherwise

If TRUE, data for B and NB are not required; however, the arguments must still appear in the calling sequence.

HIDENT   Logical variable:

TRUE   H is an identity matrix

FALSE   Otherwise

If TRUE, data for H and NH are not required; however; the arguments must still appear in the calling sequence.

IOP   Print option parameter:

IOP = 0   No printed output

Otherwise   Print input and computed results

DUMMY   Vector of working space for computations, dimensioned at least $2n^2$

c. Output Arguments

C   One-dimensional array of dimension at least $(n + 1)n^2$.
Upon normal return, C contains the matrices C_i,
$(i = 0,1,\ldots,n)$, packed by columns as $\left[C_0, C_1,\ldots, C_n\right]$.

NC   Two-dimensional vector holding, upon normal return, the
number of rows and columns of C:

NC(1) = n

NC(2) = nx(n + 1)

HCB   One-dimensional array of dimension at least $n \times m \times r$.
This array is not required if both BIDENT and HIDENT are
TRUE. However, the argument HCB must still appear in
the calling sequence. Otherwise, upon normal return,
HCB contains the matrices HC_iB, $(i = 0,1,\ldots,n-1)$,
packed by columns as $\left[HC_0B, HC_1B, \ldots, HC_{n-1}B\right]$.

NHCB   Two-dimensional vector holding, upon normal return, the
number of rows and columns of HCB:

NHCB(1) = m

NHCB(2) = n $\times$ r

Not required if both HIDENT and BIDENT are TRUE, but
must still appear as an argument of the calling sequence.

D   One-dimensional array of dimension at least n. Upon normal
return, D contains elements d_i, $(i = 1,\ldots,n)$ of
(4-50) stored as $D(I) = d_i$, $(I = i = 1,\ldots,n)$.

ND   Two-dimensional vector holding, upon normal return, the
number of rows and columns of D:

ND(1) = n

ND(2) = 1

d. COMMON Blocks

None

e. Error Messages

None

f. Subroutines Employed by LEVIER

LNCNT, PRNT, UNITY, EQUATE, MULT, TRCE, SCALE, ADD

g. Subroutines Employing LEVIER

None

h. Concluding Remarks

None

REFERENCES

4-1. B. N. Parlett and C. Reinsch, "Balancing a Matrix for Calculation of Eigenvalues and Eigenvectors," Numer. Math., Bd. 13, Heft 4, 1969, pp. 293-304.

4-2. R. S. Martin and J. H. Wilkinson, "Similarity Reduction of a General Matrix to Hessenberg Form," Numer. Math., Bd. 12, Heft 5, 1968, pp. 349-368.

4-3. R. S. Martin, G. Peters, and J. H. Wilkinson, "The QR Algorithm for Real Hessenberg Matrices," Numer. Math., Bd. 14, Heft 3, 1970, pp. 219-231.

4-4. J. H. Wilkinson and C. Reinsch, Handbook for Automatic Computation, Volume II - Linear Algebra, Springer-Verlag, 1971.

4-5. E. S. Armstrong, "Digital Explicit Model Following With Unstable Model Dynamics," AIAA Paper No. 74-888, AIAA Mechanics and Control of Flight Conference, Aug. 1974.

4-6. R. A. Smith, "Matrix Equation XA + BX = C," SIAM J. Appl. Math., Vol. 16, no. 2, Mar. 1968, pp. 198-201.

4-7. R. H. Bartels and G. W. Stewart, "Algorithm 432 - Solution of the Matrix Equation AX + XB = C," Commun. ACM, vol. 15, no. 9, Sept. 1972.

4-8. C. Källström, Computing Exp(A) and $\int$ Exp(As)ds, Rep. 7309, Lund Inst. Technol. (Sweden), Mar. 1973.

4-9. R. C. Ward, Numerical Computation of the Matrix Exponential With Accuracy Estimate, UCCND-CSD-24, Nov. 1975.

4-10. H. Kwakernaak and R. Sivan, Linear Optimal Control Systems, John Wiley and Sons, Inc., c.1972.

4-11. C. A. Desoer, Notes for a Second Course on Linear Systems, Van Nostrand Reinhold Co., c.1970.

FIVE/ PROGRAMS FOR CONTROL LAW DESIGN

I/ INTRODUCTION

This chapter describes ORACLS programs for designing linear state variable feedback control laws. The subroutines primarily employ Linear-Quadratic-Gaussian (LQG) methodology. Subroutines are given to implement both the continuous and discrete system versions of LQG theory. Included are algorithms for the transient regulator, sampled-data regulator, asymptotic regulator and Kalman-Bucy filter, and both the explicit and implicit versions of model following. Two distinct methods are available for solving the steady-state (algebraic) Riccati equation in both continuous and discrete forms. An algorithm is also given for single input eigenvalue placement.

II/ SUBROUTINE DESCRIPTIONS

A/ Evaluate Sampled-Data Regulator Coefficients (SAMPL)

1. PURPOSE

Subroutine SAMPL computes the matrix functions,

$$\tilde{Q}(T) = \int_0^T e^{A'\tau} Q e^{A\tau} \, d\tau \tag{5-1}$$

$$W(T) = 2 \int_0^T e^{A'\tau} Q H(\tau,0) \, d\tau \tag{5-2}$$

$$\tilde{R}(T) = \int_0^T \left[R + H'(\tau,0) \, Q \, H(\tau,0) \right] d\tau \tag{5-3}$$

where

$$H(t,0) = \int_0^t e^{A\tau} B \, d\tau \tag{5-4}$$

for constant real matrices A, B, $Q = Q' \geqq 0$, $R' = R > 0$, and scalar $T > 0$. The dimensions of A and B are $n \times n$ and $n \times r$ $(r \leqq n)$, respectively. Other matrices have compatible dimensions. The program has the option of computing $\tilde{Q}(T)$ without evaluating $W(T)$ and $\tilde{R}(T)$.

The matrix $\tilde{Q}$ is the reconstructibility Gramian [5-1] for the system

$$
\left.
\begin{aligned}
\dot{x}(t) &= A \, x(t) \\[2mm]
y(t) &= D \, x(t) \\[2mm]
Q &= D'D
\end{aligned}
\right\} \qquad (0 \leqq t \leqq T) \tag{5-5}
$$

The set of matrices $\tilde{Q}$, W, and $\tilde{R}$ occur naturally in the optimal sampled-data linear regulator problem described in [5-2] and [5-3]. Given the time-invariant linear system,

$$\dot{x}(t) = A \, x(t) + B \, u(t) \tag{5-6}$$

the optimal sampled-data regulator problem (OSR) occurs when $u(t)$ $(0 \leqq t \leqq t_f < \infty)$ is required to minimize

$$J = x'(t_f) \, S \, x(t_f) + \int_0^{t_f} \left[x'(\tau) \, Q \, x(\tau) + u'(\tau) \, R \, u(\tau) \right] d\tau \tag{5-7}$$

subject to the restrictions that $u(t)$ be constant over subintervals $(t_i, t_{i+1}]$ $(i = 0, 1, \ldots, N-1)$, $0 < t_0 < t_1 < t_2 \ldots < t_N = t_f$ within the interval $(0, t_f]$, and $S = S' \geqq 0$. The OSR problem transforms

84

directly into an optimal discrete regulator problem in which $u(t_i)$ $(i = 0,1,\ldots,N-1)$ is chosen to minimize

$$J = x'(t_f)\ S\ x(t_f) + \sum_{i=0}^{N-1} \left[x'(t_i)\ \tilde{Q}(\Delta t_i)\ x(t_i) \right.$$

$$\left. + x'(t_i)\ W(\Delta t_i)\ u(t_i) + u'(t_i)\ \tilde{R}(\Delta t_i)\ u(t_i) \right] \tag{5-8}$$

with

$$\Delta t_i = t_{i+1} - t_i \tag{5-9}$$

and

$$x(t_{i+1}) = e^{A\Delta t_i}\ x(t_i) + H(\Delta t_i,0)\ u(t_i) \tag{5-10}$$

The subroutine SAMPL computes the weighting matrices for the OSR problem for the case $\Delta t_i = T$.

Computation is based on the finite-series algorithm described in [5-4]. The matrix AT is scaled by $1/2^k$ where k is a positive integer chosen so that the scaled matrix has eigenvalues within the unit circle in the complex plane. The algorithm [5-4] is applied to the scaled matrix until convergence occurs. Numerically, convergence is assumed to have occurred in each series when the improvement in the element of largest magnitude (measured relatively if the magnitude is less than unity, and absolutely otherwise) is less than the parameter SERCV found in the COMMON block CONV of subroutine RDTITL. Afterward, the desired OSR weighting matrices are reconstructed from the results obtained using the scaled AT matrix.

2. USAGE

a. Calling Sequence

CALL SAMPL(A,NA,B,NB,Q,NQ,R,NR,W,NW,T,IOP,DUMMY)

b. Input Arguments

A, B, Q, R   Matrices packed by columns in one-dimensional arrays.

A and B are not destroyed upon return, but Q and R are. If only $\tilde{Q}$ is to be computed, data for B and R need not be input, but the related arguments should still appear in the calling sequence.

NA, NB, NQ, NR   Two-dimensional vectors giving the number of rows and columns of respective matrices; for example,

NA(1) = Number of rows of A

NA(2) = Number of columns of A

Not destroyed upon normal return

T   Positive scalar parameter

IOP   Two-dimensional option vector:

IOP(1) = 0   Do not print results of computation.

Otherwise   Print input and computed results.

IOP(2) = 0   Solve for the reconstructibility Gramian only.

Otherwise   Solve for the complete set of OSR weighting matrices.

DUMMY   Vector of working space for computations with dimension at least:

$4n^2$ for   IOP(2) = 0

$7n^2$ for   IOP(2) ≠ 0

c. Output Arguments

Q   Upon normal return, the matrix $\tilde{Q}(T)$ is stored in Q and packed by columns in the one-dimensional array.

W   Matrix packed by columns in a one-dimensional array with dimension at least nr. Upon normal return, W

contains the matrix $W(T)$ if computation is required.
If the computation of $W(T)$ is not required, W is
not needed, but the argument should still appear in
the calling sequence.

NW   Two-dimensional vector giving, upon normal return, the
number of rows and columns of $W(T)$:

NW(1) = NA(1)

NW(2) = NB(2)

R   Upon normal return, the matrix $\tilde{R}(T)$ is stored in R
and packed by columns in the one-dimensional array.

d. COMMON Block

CONV

e. Error Message

If k reaches 1000 in the scaling of AT, the message "ERROR IN SAMPL,
K = 1000" is printed, and the program is returned to the calling point.

f. Subroutines Employed by SAMPL

PRNT, LNCNT, NORMS, SCALE, EQUATE, MULT, TRANP, ADD, MAXEL, EXPSER, EXPINT

g. Subroutines Employing SAMPL

None

h. Concluding Remarks

The variance matrix for the state of the time-invariant linear system

$$\dot{x}(t) = A\ x(t) + B\ \eta(t) \tag{5-11}$$

driven by white noise η with intensity V is given at time T by

$$G(T) = e^{AT}\ G(0)\ e^{A'T} + \int_0^T e^{A\tau} BVB' e^{A'\tau}\ d\tau \tag{5-12}$$

and can be computed through the matrix exponential and SAMPL subroutines
of ORACLS. For the second term in $G(T)$, use SAMPL and compute the
reconstructibility Gramian with A replaced by A' and $Q = BVB'$.

87

B/ Eliminate Performance Index Cross-Products (PREFIL)

1. PURPOSE

Subroutine PREFIL computes an $r \times n$ $(r \leqq n)$ matrix F which, when used in the vector equation,

$$u = -Fx + v \tag{5-13}$$

eliminates the cross-product term in the quadratic scalar function,

$$x'Qx + x'Wu + u'Ru \tag{5-14}$$

where $Q = Q' \geqq 0$, W, and $R = R' > 0$ are constant matrices. Specifically,

$$F = R^{-1} \frac{W'}{2} \tag{5-15}$$

and, after substitution, the quadratic function becomes

$$x'\tilde{Q}x + v'Rv \tag{5-16}$$

where

$$\tilde{Q} = Q - \frac{W}{2} F \tag{5-17}$$

If the transformation is also applied to a continuous or discrete linear time-invariant system,

Continuous

$$\dot{x} = Ax + Bu \tag{5-18}$$

Discrete

$$x(i+1) = A \, x(i) + B \, u(i) \tag{5-19}$$

the closed-loop response matrix is

$$\tilde{A} = A - BF \tag{5-20}$$

Options are provided within PREFIL to compute both $\tilde{Q}$ and $\tilde{A}$ in addition to F.

2. USAGE

a. Calling Sequence

CALL PREFIL(A,NA,B,NB,Q,NQ,W,NW,R,NR,F,NF,IOP,DUMMY)

b. Input Arguments

A, B, Q, W, R	Matrices packed by columns in one-dimensional arrays. Inputs B and R are not destroyed upon normal return. If only the matrix F is required, inputs A, B, and Q are not required but must still appear in the calling sequence. Similarly, if only F and $\tilde{Q}$ are required, input B is not required, but B must still appear in the calling sequence.
NA, NB, NQ, NW, NR	Two-dimensional vectors giving the number of rows and columns in the respective matrices; for example,
	NA(1) = Number of rows of A
	NA(2) = Number of columns of A
	Not destroyed upon return
IOP	Three-dimensional option vector:
	IOP(1) = 0 Do not print results.
	Otherwise Print input data, F, and $(\tilde{Q}, \tilde{A})$ when computed.
	IOP(2) = 0 Do not compute $\tilde{Q}$.
	Otherwise Compute F and $\tilde{Q}$ and return.
	IOP(3) = 0 Do not compute $\tilde{A}$.
	Otherwise Compute F and $\tilde{A}$ and return.
DUMMY	Vector of working space for computations, dimensioned at least $n^2 + r$ where n and r are the order of A and R, respectively

c. Output Arguments

A   If $\tilde{A}$ is computed, the A array contains, upon normal
return, the matrix $\tilde{A}$ packed by columns.

Q   If $\tilde{Q}$ is computed, the Q array contains, upon normal
return, the matrix $\tilde{Q}$ packed by columns.

F   Upon normal return, the matrix F packed by columns in a
one-dimensional array of dimension at least nr

NF   Upon normal return, a two-dimensional vector giving the
number of rows and columns of F:

$$NF(1) = NR(1)$$

$$NF(2) = NA(1)$$

d. COMMON Blocks

None

e. Error Message

If the matrix R is found not to be symmetric positive definite by the
subroutine SYMPDS, the message "IN PREFIL, THE MATRIX R IS NOT SYMMETRIC
POSITIVE DEFINITE" is printed, and the program is returned to the calling
point.

f. Subroutines Employed by PREFIL

PRNT, LNCNT, TRANP, SCALE, EQUATE, SYMPDS, MULT, SUBT

g. Subroutine Employing PREFIL

IMPMDFL

h. Concluding Remarks

Subroutines which follow provide solution algorithms for LQG problems
having a quadratic performance index without cross-product terms. Problems
with such terms may be transformed into equivalent problems without cross
products by performing the control variable transformation:

$$u = -Fx + v \qquad\qquad (5\text{-}21)$$

where u is the original control, x is the state vector, and v is the new control with F computed from PREFIL.

C/ Stabilize Continuous System (CSTAB)

1. PURPOSE

Subroutine CSTAB computes a gain matrix F which, when used in the control law

$$u = -Fx \qquad\qquad (5\text{-}22)$$

and applied to the stabilizable linear time-invariant continuous system,

$$\dot{x} = Ax + Bu \qquad\qquad (5\text{-}23)$$

produces a closed-loop response matrix $(A - BF)$ whose eigenvalues lie in the complex left half-plane. The primary use for CSTAB in ORACLS is to generate a stabilizing gain matrix for initializing the quasilinearization method for solving the continuous steady-state Riccati equation [5-5]. The matrix $(A - BF)$ computed here has some interesting root-locus properties [5-6] which may make the control law applicable in other areas.

Computation follows the method described in [5-7]. The matrices A and B are of dimension $n \times n$ and $n \times r$ $(r \leqq n)$, respectively. A scalar parameter $\beta > 0$ is first selected so that the matrix

$$\tilde{A} = -(A + \beta I) \qquad\qquad (5\text{-}24)$$

has eigenvalues in the complex left half-plane, where I is an $n \times n$ identity matrix. The $\tilde{A}$ matrix is used to form a Liapunov equation

$$\tilde{A}Z + Z\tilde{A}' = -2BB' \qquad (5\text{-}25)$$

whose solution Z is used to compute F through

$$F = B'Z^+ \qquad (5\text{-}26)$$

where $+$ denotes matrix pseudoinverse. It is shown in [5-6] that

$$Re(\lambda^{A-BF}) = -\beta \qquad (5\text{-}27)$$

where λ^{A-BF} is any controllable eigenvalue of the (A,B) system. The option is provided to input β directly or to have β computed internally as

$$\beta = s\left[\max_i \left|Re\left(\lambda_i^A\right)\right| + 0.001\right] \qquad (i = 1,2,\ldots,n) \qquad (5\text{-}28)$$

where s is an input scale factor and λ_i^A are the eigenvalues of A. If the eigenvalue computation in CSTAB fails, β is set to

$$\beta = 2\|A\| \qquad (5\text{-}29)$$

using the ℓ_1 matrix norm (see subroutine NORMS). The Liapunov equation for Z can be solved using either subroutine BILIN or subroutine BARSTW. The pseudoinverse of Z is computed through subroutine SNVDEC. Computational parameters for BILIN, BARSTW, and SNVDEC are provided internally through the COMMON block TOL found in subroutine RDTITL.

2. USAGE

a. Calling Sequence

CALL CSTAB(A,NA,B,NB,F,NF,IOP,SCLE,DUMMY)

b. Input Arguments

A, B   Matrices packed by columns in one-dimensional arrays; not
            destroyed upon normal return

NA, NB   Two-dimensional vectors giving the number of rows and columns
            of the respective matrices; for example,

            NA(1) = Number of rows of A

            NA(2) = Number of columns of A

            Not destroyed upon normal return

IOP   Three-dimensional option vector:

            IOP(1) = 0   Do not print results.

            Otherwise   Print input, β, F, and eigenvalues of (A-BF).

            IOP(2) = 0   Do not compute parameter β but use β = SCLE.

            Otherwise   Compute β using s = SCLE.

            IOP(3) = 0   Use the BARSTW algorithm to solve the
                          Z equation.

            Otherwise   Use BILIN.

SCLE   Parameter used to define β

DUMMY   Vector of working space for computations of dimension at least:

            $2n^2 + 2(n + 1)^2$ if IOP(3) = 0

            $6n^2 + 2n$ if IOP(3) $\neq$ 0

c. Output Arguments

F   Matrix packed by columns into a one-dimensional array of
            dimension at least nr. Upon normal return, F contains
            $B'Z^+$.

NF   Two-dimensional vector holding, upon normal return, the number
            of rows and columns of F:

            NF(1) = NB(2)

            NF(2) = NA(1)

93

d. COMMON Block

TOL

e. Error Messages

(1) In the β computation section, if the eigenvalue computation for A
fails, the message "IN CSTAB, THE SUBROUTINE EIGEN FAILED TO
DETERMINE THE _____ EIGENVALUE FOR THE MATRIX A AFTER
30 ITERATIONS" is printed, and the computation continues with
$\beta = 2\|A\|$.

(2) If SNVDEC fails to compute the singular values of Z, the message
"IN CSTAB, SNVDEC HAS FAILED TO CONVERGE TO THE _____ SINGULAR
VALUE AFTER 30 ITERATIONS" is printed, and the program is returned
to the calling point.

(3) If a singular value of Z is greater than but close to the ZTEST
value used (see SNVDEC and TOL), the message "IN CSTAB, THE MATRIX
SUBMITTED TO SNVDEC USING ZTEST = _____ IS CLOSE TO A MATRIX OF
LOWER RANK/IF THE ACCURACY IAC IS REDUCED THE RANK MAY ALSO BE
REDUCED/CURRENT RANK = _____" is printed along with the singular
values of Z after which computation continues.

f. Subroutines Employed by CSTAB

EQUATE, EIGEN, LNCNT, TRANP, MULT, SCALE, BARSTW, BILIN, SNVDEC, PRNT,
SUBT, NORMS

g. Subroutines Employing CSTAB

DSTAB, ASYMREG

h. Concluding Remarks

None

D/ Stabilize Discrete System (DSTAB)

1. PURPOSE

Subroutine DSTAB computes a gain matrix F which, when used in the
control law,

$$u = -Fx \qquad (5\text{-}30)$$

and applied to the stabilizable linear time-invariant discrete system,

$$x(i+1) = A\ x(i) + B\ u(i) \qquad (5\text{-}31)$$

produces a closed-loop response matrix $(A - BF)$ whose eigenvalues lie inside the unit circle of the complex plane. The primary use for DSTAB in ORACLS is to generate a stabilizing gain matrix for initializing the quasilinearization method for solving the discrete steady-state Riccati equation [5-8]. The matrix $(A - BF)$ computed here has some interesting root-locus properties [5-6] which may make the control law applicable in other areas.

Computation follows the method described in [5-9]. The matrices A and B are of dimension $n \times n$ and $n \times r$ $(r \leqq n)$, respectively. A scalar α is first selected so that

$$0 < \alpha < \min\left(1,\ \min_i \left|\lambda_i^A\right|\right) \qquad (i = 1,2,\ldots,n) \qquad (5\text{-}32)$$

where $\left|\lambda_i^A\right|$ denotes the nonzero complex moduli of eigenvalues of A. Next, solve the matrix equation,

$$AZA' = \alpha^2 Z + 2BB' \qquad (5\text{-}33)$$

for $Z = Z' \geqq 0$, and by assuming the controllable eigenvalues of A are nonzero, the stabilizing gain is given by

$$F = B'(Z + BB')^+ A \qquad (5\text{-}34)$$

where $+$ denotes matrix pseudoinverse. If any controllable eigenvalue of A is zero or unknown, it can be made nonzero by applying to the system a prefilter with gain G so that the controllable eigenvalues of (A - BG) are strictly in the complex left half-plane. Such a gain

can be computed from CSTAB. Afterward, DSTAB is applied to the

(A - BG,B) pair to compute a digital stabilizing gain $\tilde{F}$ with the net

stabilizing gain F for the original system given by

$$F = G + \tilde{F} \tag{5-35}$$

In DSTAB, the matrix equation (5-33) is transformed into

$$\tilde{A}Z + Z\tilde{A}' = \tilde{B}\tilde{B}' \tag{5-36}$$

with

$$\tilde{A} = (\alpha I + A)^{-1}(A - \alpha I) \tag{5-37}$$

$$\tilde{B} = 2(\alpha I + A)^{-1}B \tag{5-38}$$

and solved by subroutine BARSTW. The symbol I denotes an n × n

identity matrix.

The option is provided to input α directly or to have α computed

internally as

$$\alpha = s \ \min \left(1, \ \min_{i} \ \left|\lambda_i^{A-BG}\right|\right) \qquad (i = 1,2,\ldots,n) \tag{5-39}$$

where $\left|\lambda_i^{A-BG}\right|$ are nonzero complex moduli of eigenvalues of (A - BG),

0 < s < 1 is input scale factor, and G is a gain computed, if

requested, internally from CSTAB to cause the controllable eigenvalues of

(A - BG) to be nonsingular. The pseudoinverse is computed through

SNVDEC. Computational parameters for SNVDEC and BARSTW are provided

internally through the COMMON block TOL found in subroutine RDTITL.

2. USAGE

a. Calling Sequence

CALL DSTAB(A,NA,B,NB,F,NF,SING,IOP,SCLE,DUMMY)

b. Input Arguments

A, B   Matrices packed by columns in one-dimensional arrays; not
               destroyed upon normal return

NA, NB   Two-dimensional vectors giving the number of rows and columns
               in the respective matrices; for example,

               $NA(1)$ = Number of rows of   A

               $NA(2)$ = Number of columns of   A

               Not destroyed upon normal return

SING   Logical variable:

               FALSE   If the controllable eigenvalues of   A   are known to
                                be nonzero

               TRUE   Otherwise, and a gain   G   will automatically
                                be computed (using CSTAB) before proceeding to the
                                digital algorithm

IOP   Two-dimensional option vector:

               $IOP(1)$ = 0   Do not print results.

               Otherwise   Print input,   F, α, eigenvalues of   A - BF,
                                and their magnitudes.

               $IOP(2)$ = 0   Do not compute parameter α but use α = SCLE.

               Otherwise   Compute α by using   s = SCLE.

SCLE   Parameter used to define α

DUMMY   Vector of working space for computations, dimensioned at least
               $4n^2 + 2(n + 1)^2$

c. Output Arguments

F   Matrix packed by columns into a one-dimensional array of
               dimension at least   nr.   Upon normal return,   F   contains
               $B'(Z + BB')^{+}A$.

NF   Two-dimensional vector holding, upon normal return, the number

of rows and columns in F:

NF(1) = NB(2)

NF(2) = NA(1)

d. COMMON Block

TOL

e. Error Messages

(1) In the α computation, if the eigenvalue computation for (A - BG)
fails, the message "IN DSTAB, THE PROGRAM EIGEN FAILED TO DETERMINE
THE _____ EIGENVALUE FOR THE MATRIX A - BG AFTER 30 ITERATIONS"
is printed followed by the matrices (A - BG) and G, if
SING = TRUE, and the program is returned to the calling point.

(2) If the matrix A + αI (or (A - BG) + αI) is found to be singular,
the message "IN DSTAB, GELIM HAS FOUND THE MATRIX A + (ALPHA)I
(or (A-BG) + (ALPHA)I) SINGULAR" is printed followed by A, G,
and α, and the program is returned to the calling point.

(3) If SNVDEC fails to compute the singular values of (Z + BB'), the
message "IN DSTAB, SNVDEC HAS FAILED TO CONVERGE TO THE _____
SINGULAR VALUE AFTER 30 ITERATIONS" is printed, and the program
is returned to the calling point.

(4) If a singular value of (Z + BB') is greater than, but close to the
ZTEST value used (see SNVDEC and TOL), the message "IN DSTAB, THE
MATRIX SUBMITTED TO SNVDEC, USING ZTEST = _____, IS CLOSE TO A
MATRIX OF LOWER RANK/IF THE ACCURACY IAC IS REDUCED THE RANK MAY
ALSO BE REDUCED/CURRENT RANK = _____" is printed along with the
singular values of (Z + BB') after which computation continues.

(5) If the eigenvalue computation for (A - BF) fails, the message "IN
DSTAB, THE PROGRAM EIGEN VAILED TO DETERMINE THE _____ EIGENVALUE

FOR THE A-BF MATRIX AFTER 30 ITERATIONS" is printed along with
the computed eigenvalues, and the program is returned to the
calling point.

f. Subroutines Employed by DSTAB

CSTAB, MULT, SUBT, EQUATE, EIGEN, GELIM, LNCNT, PRNT, TRANP, MULT,

SCALE, BARSTW, ADD, SNVDEC, JUXTC

g. Subroutine Employing DSTAB

ASYMREG

h. Concluding Remarks

None

E/ Digital Transient Quadratic Regulator (DISCREG)

1. PURPOSE

Subroutine DISCREG solves the time-invariant discrete-time linear optimal
output regulator problem with noise-free measurements. Given the digital
linear system

$$x(i+1) = A x(i) + B u(i) + w(i) \qquad (5\text{-}40)$$

where $x(0) = x_0$ is given, A and B are constant matrices of
dimension $n \times n$ and $n \times r$ $(r \leqq n)$, respectively, and $w(i)$
$(i = 0,1,\ldots,N\text{-}1)$ is a sequence of uncorrelated zero-mean stochastic
variables with variance matrices $V(i)$. Outputs, or controlled variables,
are of the form

$$y(i) = H x(i) \qquad (5\text{-}41)$$

where H is a constant $m \times n$ $(m \leqq n)$ matrix. The considered optimal
regulator problem is to find the control sequence $u(i)$ which minimizes

$$J = E \left\{ \sum_{i=0}^{N-1} \left[y'(i+1) \ Q \ y(i+1) + u'(i) \ R \ u(i) \right] \right.$$

$$\left. + \ x'(N) \ P_N \ x(N) \right\} \qquad\qquad (5\text{-}42)$$

with $Q = Q' \geqq 0$, $P_N = P_N' \geqq 0$, and $R = R' > 0$. The symbol E denotes expected value. The solution is given by [5-10] as

$$\left. \begin{aligned} u(i) &= -F(i) \ x(i) \\[2mm] F(i) &= \left[R + B' \ P(i+1) \ B \right]^{-1} B' \ P(i+1) \ A \\[2mm] P(i) &= \phi'(i) \ P(i+1) \ \phi(i) + F'(i) \ R \ F(i) + H'QH \\[2mm] \phi(i) &= A - BF(i) \end{aligned} \right\} \quad (i = 0,1,\ldots,N\text{-}1) \quad (5\text{-}43)$$

where $P(N) = P_N$. The minimal value of the criterion function is

$$J = x'(0) \ P(0) \ x(0) + \sum_{j=0}^{N-1} \text{tr} \left[V(j) \ P(j+1) \right] \qquad (5\text{-}44)$$

where tr denotes the trace of a matrix.

The program DISCREG repetitively evaluates the solution equations from an input value N to stage zero or until $P(i)$ converges to a steady-state value, whichever occurs first. Numerically, convergence is assumed to have occurred when the improvement in the element of largest magnitude (measured relatively if the magnitude is less than unity, and absolutely otherwise) is less than the parameter RICTCV found in COMMON block CONV of subroutine RDTITL. The program DISCREG does not evaluate the optimal value of the criterion function or print a response trajectory. The subroutine SNVDEC is used to evaluate the matrix inverse in the $F(i)$ equation to allow the option of using nonnegative definite (instead of strictly positive definite) matrices R. Parameters for SNVDEC are set internally by use of the COMMON block TOL of RDTITL.

2. USAGE

a. Calling Sequence

CALL DISCREG(A,NA,B,NB,H,NH,Q,NQ,R,NR,F,NF,P,NP,IOP,IDENT,DUMMY)

b. Input Arguments

A, B, H, Q, R	Matrices packed by columns in one-dimensional arrays. Upon normal return, all except Q are not destroyed.
NA, NB, NH, NQ, NR	Two-dimensional vectors giving the number of rows and columns of the respective matrices; for example,
	NA(1) = Number of rows of A
	NA(2) = Number of columns of A
	Not destroyed upon normal return except for NQ = NA
P	Symmetric nonnegative definite matrix packed by columns in a one-dimensional array. On input, P contains the matrix P_N. Afterward, intermediate values of $P(i)$ are stored in the array P.
NP	Two-dimensional vector giving the number of rows and columns of P:
	NP(1) = Number of rows in P
	NP(2) = Number of columns in P
	Not destroyed upon normal return

IOP   Three-dimensional option vector:

                  IOP(1) = 0   Do not print results.

                  Otherwise   Print input, and F(i) and P(i) at
                                stage zero or steady state, whichever
                                occurs first.

                  IOP(2) = 0   Do not print at intermediate states between
                                initial and final states.

                  Otherwise   Print stage count and values of F(i)
                                and P(i), regardless of printing speci-
                                fied by IOP(1).

                  IOP(3) is terminal stage N for the optimal regulator
                  problem.

IDENT   Logical variable:

                  TRUE   If H is an identity matrix

                  FALSE   Otherwise

                  For IDENT = TRUE, no data need be input for H, but the
                  related arguments should still appear in the calling
                  sequence.

DUMMY   Vector of working space for computations of dimension at
                  least $4n^2$

c. Output Arguments

Q   Upon normal return, Q is replaced by H'QH.   The
                  array Q must be dimensioned at least n^2.

F   Matrix packed by columns in a one-dimensional array of
                  dimension at least nr. Upon normal return, F contains
                  the value of F(i) at stage zero or the stage at which
                  numerical steady-state convergence occurs.

NF   Two-dimensional matrix giving, upon normal return, the number of rows and columns of F:

$$NF(1) = NB(2)$$
$$NF(2) = NA(1)$$

P   Upon normal return, P contains the value of $P(i)$ at stage zero or the stage at which steady-state convergence occurs.

d. COMMON Blocks

TOL, CONV

e. Error Messages

(1) If SNVDEC fails to compute the singular values of any $[R + B' \ P(i) \ B]$, the message "IN DISCREG, SNVDEC HAS FAILED TO CONVERGE TO THE _____ SINGULAR VALUE AFTER 30 ITERATIONS" is printed, and the program is returned to the calling point.

(2) If a singular value of $[R + B' \ P(i) \ B]$ is greater than but close to the ZTEST value used (see SNVDEC and TOL), the message "IN DISCREG, THE MATRIX SUBMITTED TO SNVDEC USING ZTEST = _____ IS CLOSE TO A MATRIX OF LOWER RANK/IF THE ACCURACY IAC IS REDUCED THE RANK MAY ALSO BE REDUCED/CURRENT RANK = _____" is printed along with the singular values of $[R + B'P(i)B]$ after which computation continues.

f. Subroutines Employed by DISCREG

LNCNT, PRNT, MULT, TRANP, EQUATE, SNVDEC, ADD, SUBT, MAXEL

g. Subroutine Employing DISCREG

ASYMREG

h. Concluding Remarks

Of course, if $w(i) = 0$ ($i = 0, 1, \ldots$), the algorithm in DISCREG generates the solution to the discrete deterministic optimal linear output regulator problem.

F/ Continuous Transient Quadratic Regulator (CNTNREG)

1. PURPOSE

Subroutine CNTNREG solves the time-invariant continuous-time linear optimal output regulator problem with noise-free measurements. Given the continuous linear system,

$$\dot{x}(t) = A\ x(t) + B\ u(t) + w(t) \qquad (5\text{-}45)$$

where $x(0) = x_0$ is given, A and B are constant matrices of dimension $n \times n$ and $n \times r$ $(r \leqq n)$, respectively, and $w(t)$ is zero-mean Gaussian white noise with intensity $V(t)$. Outputs, or controlled variables, are modeled as

$$y(t) = H\ x(t) \qquad (5\text{-}46)$$

where H is a constant $m \times m$ $(m \leqq n)$ matrix. The considered optimal regulator problem is to find the control function $u(t)$ which minimizes

$$J = E\left\{ \int_0^{t_1} \left[y'(t)\ Q\ y(t) + u'(t)\ R\ u(t) \right] dt + x'(t_1)\ P_1\ x(t_1) \right\} \qquad (5\text{-}47)$$

where $Q = Q' \geqq 0$, $P_1 = P_1' \geqq 0$, and $R = R' > 0$. The symbol E denotes expected value. The solution is given from [5-1] as

$$u(t) = -F(t)\ x(t) \qquad (5\text{-}48)$$

$$F(t) = R^{-1}B'\ P(t) \qquad (5\text{-}49)$$

$$\frac{-dP(t)}{dt} = H'QH + A'\ P(t) + P(t)\ A - P(t)\ BR^{-1}B'\ P(t) \qquad (5\text{-}50)$$

where $P(t_1) = P_1$. The minimal value of the criterion function is

$$x'(0)\ P(0)\ x(0) + \int_0^{t_1} \text{tr}\left[P(t)\ V(t) \right] dt \qquad (5\text{-}51)$$

where tr denotes the trace of a matrix.

The computational algorithm used in CNTNREG to solve the Riccati equation is due to Vaughan [5-11] as described by Kwakernaak and Sivan [5-1]. First, form

$$Z = \begin{bmatrix} A & -BR^{-1}B' \\ -H'QH & -A' \end{bmatrix} \qquad (5-52)$$

The Z matrix has the property that if λ is an eigenvalue, then $-\lambda$ is also an eigenvalue. By assuming that Z has linearly independent eigenvectors, there exists a matrix W such that

$$Z = W \begin{pmatrix} \Lambda & 0 \\ 0 & -\Lambda \end{pmatrix} W^{-1} \qquad (5-53)$$

where Λ is a diagonal matrix constructed as follows. If an eigenvalue λ of Z has $Re(\lambda) > 0$, it is a diagonal element of Λ, and $-\lambda$ is automatically placed in the corresponding location of $-\Lambda$. If $Re(\lambda) = 0$, one of the pair $(\lambda, -\lambda)$ is arbitrarily assigned to Λ and the other to $-\Lambda$. The matrix W is composed of eigenvectors of Z; the ith column vector of W is the eigenvector of Z corresponding to the eigenvalue in the ith diagonal position of $diag(\Lambda, -\Lambda)$. In practice, it is noted that if λ is a complex eigenvalue of Z with eigenvector ν, then their complex conjugates $\bar{\lambda}$ and $\bar{\nu}$ are also an eigenpair and $\left[Re(\nu), Im(\nu)\right]$ is placed in W instead of $(\nu, \bar{\nu})$ in order to avoid complex arithmetic. This has the effect of making Λ block diagonal where every $(\lambda, \bar{\lambda})$ pair go together to form a 2×2 real matrix placed at the former $(\lambda, \bar{\lambda})$ entry in Λ. Next partition the $2n \times 2n$ matrix W into four $n \times n$ blocks as

$$W = \begin{bmatrix} W_{11} & W_{12} \\ W_{21} & W_{22} \end{bmatrix} \qquad (5\text{-}54)$$

The solution $P(t)$ of the Riccati equation (5-50) can be written as

$$P(t) = \left[W_{22} + W_{21}\, G(t_1 - t) \right] \left[W_{12} + W_{11}\, G(t_1 - t) \right]^{-1} \qquad (5\text{-}55)$$

with

$$G(t) = e^{-\Lambda t} S e^{-\Lambda t} \qquad (5\text{-}56)$$

and

$$S = -(W_{21} - P_1 W_{11})^{-1}(W_{22} - P_1 W_{12}) \qquad (5\text{-}57)$$

If Λ is composed of eigenvalues with strictly positive real parts, then the steady-state solution to the Riccati equation is given directly as

$$\lim_{t_1 \to \infty} P(t) = W_{22}\, W_{12}^{-1} \qquad (5\text{-}58)$$

The program CNTNREG evaluates the solution equations from an input value t_1 to time zero or until $P(t)$ converges to a steady-state value, whichever occurs first. Numerically, convergence is assumed to have occurred when the improvement in the element of largest magnitude (measured relatively if the magnitude is less than unity, and absolutely otherwise) is less than the parameter RICTCV found in the COMMON block CONV of subroutine RDTITL. CNTNREG does not evaluate the optimal value of the criterion function or print a response trajectory. Options are provided to compute directly the steady-state solution and to compute Z, W, S, Λ, and $e^{-\Lambda t}$ without computing any of the $P(t)$ values.

2. USAGE

a. Calling Sequence

CALL CNTNREG(A,NA,B,NB,H,NH,Q,NQ,R,NR,Z,W,LAMBDA,S,F,NF,P,NP,T,IOP,
IDENT,DUMMY)

b. Input Arguments

A, B, H, Q, R	Matrices packed by columns in one-dimensional arrays.
	Upon normal return, all except Q are not destroyed.
NA, NB, NH, NQ, NR	Two-dimensional vectors giving the number of rows and
	columns in the respective matrices; for example,
	NA(1) = Number of rows of A
	NA(2) = Number of columns of A
	Not destroyed upon normal return except for NQ
	replaced by NA
P	Symmetric nonnegative definite matrix packed in a one-
	dimensional array. On input, P contains the
	matrix P_1. Afterward, intermediate values of $P(t)$
	are stored in the array P.
NP	Two-dimensional vector giving the number of rows and
	columns in P:
	NP(1) = Number of rows of P
	NP(2) = Number of columns of P
	Not destroyed upon normal return
T	Two-dimensional vector:
	T(1) = Final time t_1
	T(2) = Increment of time for transient solution
	computation
	The final time t_1 must be expressed as an integer
	multiple of T(2). The vector T is not required if
	only the steady-state solution of $P(t)$ is required
	but must still appear as an argument of the calling
	sequence; not destroyed upon normal return. Set

T(1) negative if only Z, W, S, Λ, and $e^{-\Lambda T(2)}$ are required without any P(t) computation.

IOP   Three-dimensional option vector:

   IOP(1) = 0 Do not print results, nor compute $W^{-1}ZW$.

   Otherwise Print input, Z, λ^Z, W, Λ, $W^{-1}ZW$, W_{11}, W_{12}, W_{21}, W_{22}, $R^{-1}B'$, $e^{-\Lambda T(2)}$ and values of F and P at time zero or steady state, whichever comes first.

   IOP(2) = 0 Do not print at intermediate times (multiples of T(2)) between T(1) and zero or steady state.

   Otherwise Regardless of the printing specified by IOP(1), print values of P and F at intermediate times.

   The parameter IOP(2) is not required if only a steady-state solution is required.

   IOP(3) = 0 Compute transient solutions P(t) and F(t).

   Otherwise Compute only steady-state values of P and F.

IDENT   Logical variable:

   TRUE If H is an identity matrix

   FALSE Otherwise

   For IDENT = TRUE, no data need be input in H but the related arguments should still appear in the calling sequence.

DUMMY	Vector of working space for computations dimensioned at least:

$$8n^2 + 18n + nr \quad \text{for} \quad IOP(3) \neq 0$$

$$9n^2 + 17n + nr \quad \text{for} \quad IOP(3) = 0$$

c. Output Arguments

Q	Upon normal return, Q is replaced by H'QH. The array Q must be dimensioned at least n^2.
Z, W, LAMBDA, S	Matrices packed by columns in one-dimensional arrays dimensioned at least $4n^2$, $4n^2$, n^2, and n^2, respectively. Upon normal return, these arrays contain their theoretical counterparts. The array LAMBDA (Λ) is declared real.
F	Matrix packed by columns in a one-dimensional array of dimension at least nr. Upon normal return, F contains the value of $F(t)$ at time zero or the time at which steady-state convergence numerically occurs.
NF	Two-dimensional matrix giving, upon normal return, the number of rows and columns of F:

$$NF(1) = NB(2)$$

$$NF(2) = NA(1)$$

P	Upon normal return, P contains the value of $P(t)$ at time zero or the time at which steady-state convergence numerically occurs.
DUMMY	Upon normal return, the first nr elements contain the matrix $R^{-1}B'$, the next $4n^2$ contain the submatrices W_{11}, W_{21}, W_{12}, and W_{22} stored in block column form and, if a transient solution is computed, the next n^2 contain the matrix $e^{-\Lambda T(2)}$.

d. COMMON Block

CONV

e. Error Messages

(1) If the computation of $R^{-1}B'$ fails, the message "IN CNTNREG, THE SUBROUTINE SYMPDS HAS FOUND THE MATRIX R NOT SYMMETRIC POSITIVE DEFINITE" is printed, and the program is returned to the calling point.

(2) If the eigenvalue/eigenvector computation for Z fails, either the message "IN CNTNREG, THE _____ EIGENVALUE OF Z HAS NOT BEEN FOUND AFTER 30 ITERATIONS" or "IN CNTNREG, EIGEN FAILED TO COMPUTE THE _____ EIGENVECTOR OF Z" is printed, and the program is returned to the calling point.

(3) If the computation for $W^{-1}ZW$ fails, the message "IN CNTNREG, GELIM HAS FOUND THE REORDERED MATRIX W TO BE SINGULAR" is printed, and the computation continues.

(4) If the computation of S for the transient solution fails, the message "IN CNTNREG, GELIM HAS FOUND THE MATRIX W21 - P1X W11 TO BE SINGULAR" is printed, and the program is returned to the calling point.

(5) If the computation of W_{12}^{-1} fails in the steady-state case, the message "IN CNTNREG, GELIM HAS FOUND THE MATRIX W12 TO BE SINGULAR" is printed, and the program is returned to the calling point.

(6) If at any time, computation of the matrix inverse in the $P(t)$ computation fails, the message "IN CNTNREG AT TIME _____ P CANNOT BE COMPUTED DUE TO MATRIX SINGULARITY IN GELIM" is printed, and the program is returned to the calling point.

f. Subroutines Employed by CNTNREG

EQUATE, TRANP, SYMPDS, SCALE, MULT, JUXTR, EIGEN, GELIM, PRNT, SUBT, EXPSER, MAXEL, LNCNT, NULL

g. Subroutine Employing CNTNREG

ASYMREG

h. Concluding Remarks

Of course, if $w(t) = 0$ for all t, then the algorithm of CNTNREG produces the solution to the deterministic optimal linear output regulator problem. In this case, the system response $x(t)$ to the steady-state optimal control law

$$u(t) = -R^{-1}B'W_{22}W_{12}^{-1} x(t) \tag{5-59}$$

is given by

$$x(t) = W_{12}e^{-\Lambda t}W_{12}^{-1}x_0 \tag{5-60}$$

G/ Riccati Solution by Newton's Method (RICTNWT)

1. PURPOSE

Subroutine RICTNWT solves either the continuous or discrete steady-state Riccati equation by the Newton algorithms described by Kleinman [5-5] and Hewer [5-8].

For the continuous case, the algebraic Riccati equation is

$$PA + A'P + H'QH - PBR^{-1}B'P = 0 \tag{5-61}$$

where the constant matrices A, B, H, $Q = Q' \gtreqless 0$, and $R = R' > 0$ are of dimension $n \times n$, $n \times r$ ($r \lesseqgtr n$), $m \times n$ ($m \lesseqgtr n$), $m \times m$, and $r \times r$, respectively. Applying Newton's equation-solving algorithm [5-12] to solve the continuous P equation leads to the sequence

$$
\left.\begin{array}{l}
0 = \phi_k'P_k + P_k\phi_k + H'QH + F_k'RF_k \\[2mm]
\phi_k = A - BF_k \\[2mm]
F_k = R^{-1}B'P_{k-1}
\end{array}\right\} \qquad (k = 1,2,\ldots) \qquad (5\text{-}62)
$$

Kleinman [5-5] and Sandell [5-13] established that if the (A,B) pair
is stabilizable and the (A,D) pair (with D such that D'D = H'QH)
is detectable [5-1], the sequence $P_k = P_k' \geqq 0$ (k = 0,1,...) converges
to the correct $P = P' \geqq 0$ when F_1 is chosen so that (A - BF_1) is
asymptotically stable in the continuous sense. In RICTNWT, the option
is provided to solve the continuous Liapunov equation for P_k by either
of the BILIN or BARSTW subroutines.

For the discrete case, the steady-state Riccati equation can be written
as

$$
P = \phi'P\phi + F'RF + H'QH \qquad\qquad (5\text{-}63)
$$

with

$$
\phi = A - BF \qquad\qquad (5\text{-}64)
$$

$$
F = (R + B'PB)^{-1}B'PA \qquad\qquad (5\text{-}65)
$$

and A, B, H, Q, and R as previously defined. Applying Newton's
algorithm gives the sequence

$$
\left.\begin{array}{l}
P_k = \phi_k'P_k\phi_k + F_k'RF_k + H'QH \\[2mm]
F_k = (R + B'P_{k-1}B)^{-1}B'P_{k-1}A \\[2mm]
\phi_k = A - BF_k
\end{array}\right\} \qquad (k = 1,2,\ldots) \qquad (5\text{-}66)
$$

It can be established that if the (A,B) pair is stabilizable and the

112

(A,D) pair (with D such that D'D = H'QH) is detectable, the sequence
$P_k = P_k' \geq 0$ (k = 0,1,...) converges to the correct $P = P' \geq 0$ when
F_1 is chosen so that $(A - BF_1)$ is asymptotically stable in the discrete sense. In RICTNWT, the discrete Liapunov equation in P_k is
solved by subroutine SUM.

In both the continuous and discrete cases, RICTNWT assumes that an input
F_1 is provided such that $(A - F_1)$ is asymptotically stable in the
appropriate sense. If A is already stable, $F_1 = 0$ suffices. Otherwise, the subroutines CSTAB and DSTAB are available for computing
initializing F_1 matrices. RICTNWT repetitively evaluates the solution
sequence equations for up to 100 iterations or until P_k converges to P.
Numerically, convergence is assumed to have occurred when the improvement
in the element of largest magnitude (measured relatively if the magnitude
is less than unity, and absolutely otherwise) is less than the parameter
RICTCV found in COMMON block CONV of subroutine RDTITL. Parameters for
use in subroutine BARSTW are set internally through the COMMON block TOL
of RDTITL.

2. USAGE

a. Calling Sequence

CALL RICTNWT(A,NA,B,NB,H,NH,Q,NQ,R,NR,F,NF,P,NP,IOP,IDENT,DISC,FNULL,DUMMY)

b. Input Arguments

A, B, H, Q, R	Matrices packed by columns into one-dimensional arrays.
	Upon normal return, all except Q are not destroyed.
NA, NB, NH, NQ, NR	Two-dimensional vectors giving the number of rows and
	columns of the respective matrices; for example,
	NA(1) = Number of rows of A
	NA(2) = Number of columns of A

	Not destroyed upon normal return except for NQ
	replaced by NA
F	Matrix packed by columns into a one-dimensional array of size at least nr. If the matrix A is not asymptotically stable, the input F causes (A - BF) to be asymptotically stable. For A asymptotically stable, no data for F are required.
NF	Two-dimensional vector giving the number of rows and columns of F:
	NF(1) = r
	NF(2) = n
	Not required as input if data for F are not input
IOP	Three-dimensional option vector:
	IOP(1) = 0 Do not print results.
	Otherwise Print input data, H'QH, and final values of F and P.
	IOP(2) = 0 Do not print at each iteration.
	Otherwise Regardless of printing specified by IOP(1), print iteration count and value of P.
	IOP(3) = 0 Use subroutine BARSTW to solve the continuous Liapunov equation.
	Otherwise Use subroutine BILIN.
	IOP(3) is not required if the discrete Riccati equation is to be solved.
IDENT	Logical variable:

	TRUE If H is an identity matrix

TRUE   If H is an identity matrix

FALSE   Otherwise

If H is an identity matrix, no data need be input for H, but H and NH must still appear as arguments of the calling sequence.

DISC   Logical variable:

                TRUE   If the digital version is solved

                FALSE   For the continuous version

FNULL   Logical variable:

                TRUE   If $F = 0$

                FALSE   Otherwise

DUMMY   Vector of working space for computations dimensioned at least:

$5n^2$ if DISC = TRUE

$5n^2 + n(r + 4) + 2$ if DISC = FALSE and IOP(3) = 0

$7n^2 + n(r + 2)$ if DISC = FALSE and IOP(3) $\neq$ 0

c. Output Arguments

Q   Upon normal return, Q is replaced by H'QH. Q must be dimensioned at least n^2.

F   The value of F_k at the stage k of return. If convergence occurs, F contains the steady-state gain matrix.

P   Matrix packed by columns in a one-dimensional array of dimension at least n^2; the value of P_k at the stage k of return. If convergence occurs, P contains the steady-state Riccati equation solution.

NP, NF   Two-dimensional vectors giving the number of rows and columns in P and F: upon normal return,

$$NP = NA$$

$$NF(1) = r$$

$$NF(2) = n$$

d. COMMON Blocks

TOL, CONV

e. Error Messages

(1) If the computation of either R^{-1} or $(R + B'P_k B)^{-1}$ $(k = 0,1,\ldots)$ fails, the message "IN RICTNWT, A MATRIX WHICH IS NOT SYMMETRIC POSITIVE DEFINITE HAS BEEN SUBMITTED TO SYMPDS" is printed, and the program is returned to the calling point.

(2) If the iteration count exceeds 100, the message "THE SUBROUTINE RICTNWT HAS EXCEEDED 100 ITERATIONS WITHOUT CONVERGENCE" is printed, IOP(1) is set to 1, P and F are printed, and the program is returned to the calling point.

f. Subroutines Employed by RICTNWT

LNCNT, PRNT, MULT, TRANP, EQUATE, SYMPDS, SCALE, SUBT, ADD, BARSTW, BILIN, MAXEL, SUM

g. Subroutine Employing RICTNWT

ASYMREG

h. Concluding Remarks

None

H/ Asymptotic Quadratic Regulator (ASYMREG)

1. PURPOSE

Subroutine ASYMREG solves either the continuous or discrete time-invariant asymptotic linear optimal output regulator problem with noise-free measurements. For the continuous-time deterministic case, the control function $u(t)$ is chosen to minimize the criterion function,

116

$$J = \lim_{t_1 \to \infty} \int_0^{t_1} \left[y'(t)\, Q\, y(t) + u'(t)\, R\, u(t) \right] dt \qquad (5\text{-}67)$$

subject to

$$\dot{x}(t) = A\, x(t) + B\, u(t) \qquad (5\text{-}68)$$

$$y(t) = H\, x(t) \qquad (5\text{-}69)$$

where $x(0) = x_0$ is given and A, B, H, $Q = Q' \geq 0$, and $R = R' > 0$ are constant matrices of dimension $n \times n$, $n \times r$ $(r \leq n)$, $m \times n$ $(m \leq n)$, $m \times m$, and $r \times r$, respectively. If the (A,B) pair is stabilizable and the (A,D) pair (with $D'D = H'QH$) is detectable, a solution to the optimal control problem exists and is given by

$$u(t) = -F\, x(t) \qquad (5\text{-}70)$$

where

$$F = R^{-1}B'P \qquad (5\text{-}71)$$

$$PA + A'P + H'QH - PBR^{-1}B'P = 0 \qquad (5\text{-}72)$$

with the criterion function taking the value

$$x'(0)\, P\, x(0) \qquad (5\text{-}73)$$

The same control law satisfies the stochastic continuous version of the problem [5-1] in which the criterion function is

$$J = \lim_{t \to \infty} E\left[y'(t)\, Q\, y(t) + u'(t)\, R\, u(t) \right] \qquad (5\text{-}74)$$

where E is expected value, and the state equation is

$$\dot{x}(t) = A\, x(t) + B\, u(t) + w(t) \qquad (5\text{-}75)$$

117

where w(t) is a zero-mean Gaussian white noise with intensity V. The optimal value of the stochastic criterion function is here

$$\text{trace (PV)} \qquad\qquad (5\text{-}76)$$

For the deterministic discrete-time case, the control function u(i) (i = 0,1,...) is chosen to minimize the criterion function

$$J = \lim_{N\to\infty} \left\{ \sum_{i=0}^{N-1} \left[y'(i+1)\ Q\ y(i+1) + u'(i)\ R\ u(i) \right] \right\} \qquad (5\text{-}77)$$

subject to

$$x(i+1) = A\ x(i) + B\ u(i) \qquad\qquad (5\text{-}78)$$

$$y(i) = H\ x(i) \qquad\qquad (5\text{-}79)$$

where $x(0) = x_0$ is given and A, B, H, Q, and R are as previously defined. If the (A,B) pair is stabilizable and the (A,D) pair (with D'D = H'QH) is detectable, a solution to the optimal control exists and is given by

$$u(i) = -F\ x(i) \qquad\qquad (5\text{-}80)$$

where

$$F = (R + B'PB)^{-1}B'PA \qquad\qquad (5\text{-}81)$$

$$P = \phi'P\phi + F'RF + H'QH \qquad\qquad (5\text{-}82)$$

$$\phi = A - BF \qquad\qquad (5\text{-}83)$$

with the criterion function taking the value

$$x'(0)\ P\ x(0) \qquad\qquad (5\text{-}84)$$

The same control law satisfies the stochastic discrete-time version of the problem [5-1] in which the criterion function is

$$J = \lim_{N\to\infty} \frac{1}{N}\ E \left\{ \sum_{i=0}^{N-1} \left[y'(i+1)\ Q\ y(i+1) + u'(i)\ R\ u(i) \right] \right\} \qquad (5\text{-}85)$$

118

and the state equation is

$$x(i+1) = A \, x(i) + B \, u(i) + w(i) \qquad (5\text{-}86)$$

where $w(i)$ $(i = 0,1,\ldots)$ is a sequence of zero-mean stochastic variables with variance matrix V. The optimal value of the stochastic criterion function is

$$\text{trace } (PV) \qquad (5\text{-}87)$$

ASYMREG does not evaluate the optimal values of the performance criteria. Therefore, no V data are input. The option of solving the appropriate steady-state Riccati equation using either of the subroutines DISCREG, CNTNREG, or RICTNWT is provided. In addition, the residual error in the Riccati equation, the eigenvalues of P, the closed-loop response matrix $(A - BF)$, and the eigenvalues of $(A - BF)$ can be computed. If RICTNWT is selected and if the matrix A is not asymptotically stable or it is unknown whether A is asymptotically stable, the option is provided to test the relative stability of A and compute, if necessary, a stabilizing gain to initialize the Newton process.

2. USAGE

a. Calling Sequence

CALL ASYMREG(A,NA,B,NB,H,NH,Q,NQ,R,NR,F,NF,P,NP,IDENT,DISC,NEWT,STABLE, FNULL,ALPHA,IOP,DUMMY)

b. Input Arguments

A, B, H, Q, R	Matrices packed by columns into one-dimensional arrays. Upon normal return, all except Q are not destroyed.
NA, NB, NH, NQ, NR	Two-dimensional vectors giving the number of rows and columns of the respective matrices; for example,

NA(1) = Number of rows of A

NA(2) = Number of columns of A

Not destroyed upon normal return except for NQ
replaced by NA

F, NF,   F and P are matrices packed by columns into one-
P, NP   dimensional arrays of dimension at least $n r$ and n^2,
respectively.

NF and NP are the corresponding two-dimensional vectors
giving the number of rows and columns of P and F

Any input requirements depend on and are specified by
whether DISCREG, CNTNREG, or RICTNWT is employed.

IDENT   Logical variable:

TRUE   If H is an identity matrix

FALSE   Otherwise

If H is an identity matrix, no data need be input
for H, but H and NH must still appear as arguments
of the calling sequence.

DISC   Logical variables:

TRUE   If the digital version is solved

FALSE   For the continuous version

NEWT   Logical variable:

TRUE   Newton's method (RICTNWT) is used to solve the
appropriate Riccati equation.

FALSE   Either CNTNREG or DISCREG is used depending
upon the value of DISC.

STABLE   Logical variable. Data for STABLE are not required if
NEWT = FALSE, but STABLE must still appear as an argument of the calling sequence. When NEWT = TRUE:

STABLE = TRUE   If it is known that the matrix

(A - BF) computed from input data

is stable relative to an input
parameter ALPHA.

STABLE = FALSE   The matrix $(A - BF)$ is evaluated
and tested for stability relative
to ALPHA using subroutine TESTSTA.
If a stabilizing gain is required,
it is computed from subroutine
DSTAB or CSTAB.

FNULL   Logical variable; data for FNULL are not required if
NEWT = FALSE, but FNULL must still appear as an argument
of the calling sequence. When NEWT = TRUE:
FNULL = TRUE   If the input F is a null matrix
FNULL = FALSE   Otherwise

ALPHA   Scalar variable. ALPHA is not required if NEWT = FALSE
or STABLE = TRUE. Otherwise, ALPHA is used in the
asymptotic stability test of $(A - BF)$ from input data
using subroutine TESTSTA.

IOP   Five-dimensional option vector:
IOP(1), IOP(2), and IOP(3) are the first three elements
of the IOP vector in DISCREG, CNTNREG, or RICTNWT. If
CNTNREG is selected, IOP(3) is set internally to a
nonzero value.

IOP(4) = 0   Do not compute the Riccati equation
residual.
Otherwise   Compute the residual and print.

IOP(5) = 0   Compute but do not print the eigenvalues

of P, the matrix $(A - BF)$, and the
eigenvalues of $(A - BF)$.

Otherwise   Print these data after computation.

DUMMY   Vector of working space for computations dimensioned at
least:

DISC	NEWT		
TRUE	TRUE	$6n^2 + 4n + 2$	(STABLE = FALSE)
		$5n^2$	(STABLE = TRUE)
TRUE	FALSE	$4n^2$	
FALSE	FALSE	$17n^2 + n(r + 18)$	
FALSE	TRUE	$8n^2 + n(r + 1)$	(BILIN)
		$5n^2 + n(r + 4) + 2$	(BARSTW)

c. Output Arguments

Q   Upon normal return, Q is replaced by H'QH. The
dimension of Q must be at least n^2.

F   Upon normal return, F contains the steady-state gain
matrix for either the continuous or discrete problem.

NF   Two-dimensional vector giving, upon normal return, the
number of rows and columns of F:
NF(1) = r
NF(2) = n

P   Upon normal return, P contains the steady-state solution
for either the continuous or discrete Riccati equation.

NP   Two-dimensional vector giving the number of rows and columns
of P. Upon normal return, NP = NA.

STABLE   Upon normal return, STABLE = TRUE.

DUMMY   Upon normal return, the first n elements of DUMMY contain the eigenvalues of P, the next n^2 contain the

matrix $(A - BF)$ for steady-state F, and the next $2n$ contain the eigenvalues of $(A - BF)$ stored as an $n \times 2$ matrix with the real parts as first column and imaginary parts as the second. All matrices are packed by columns into one-dimensional arrays.

d. COMMON Blocks

None

e. Error Messages

(1) If the stabilizing gain computation for the continuous system fails, the message "IN ASYMREG, CSTAB HAS FAILED TO FIND A STABILIZING GAIN MATRIX (F) RELATIVE TO / ALPHA = _____" is printed, and the program is returned to the calling point.

(2) If the stabilizing gain computation for the discrete system fails, the message "IN ASYMREG, DSTAB HAS FAILED TO FIND A STABILIZING GAIN MATRIX (F) RELATIVE TO / ALPHA = _____" is printed, and the program is returned to the calling point.

(3) If the eigenvalue computation for P fails, the message "IN ASYMREG, THE _____ EIGENVALUE OF P HAS NOT BEEN COMPUTED AFTER 30 ITERATIONS" is printed, and the program continues with the computation of $(A - BF)$ and its eigenvalues.

(4) If the eigenvalue computation for $(A - BF)$ fails, the message "IN ASYMREG, THE _____ EIGENVALUE OF $A-BF$ HAS NOT BEEN COMPUTED AFTER 30 ITERATIONS" is printed, and a return is made to the calling point if no printing is required. Otherwise, the available information is printed before return.

f. Subroutines Employed by ASYMREG

MULT, SUBT, TESTSTA, CSTAB, SCALE, DISCREG, RICTNWT, TRANP, ADD, EQUATE, EIGEN, UUXTC, LNCNT, PRNT, CNTNREG, DSTAB

123

g. Subroutines Employing ASYMREG

ASYMFIL, EXPMDFL, IMPMDFL

h. Concluding Remarks

When using DISCREG to solve the steady-state discrete Riccati equation, the entry $IOP(3) = N$ must be set sufficiently large for steady-state convergence to occur.

I/ Asymptotic Kalman-Bucy Filter (ASYMFIL)

1. PURPOSE

Subroutine ASYMFIL solves either the continuous or discrete time-invariant asymptotic optimal Kalman-Bucy filter problem [5-1].

For the continuous case, the state equation is given by

$$\dot{x}(t) = A\ x(t) + G\ \tilde{n}(t) \tag{5-88}$$

with output

$$y(t) = H\ x(t) + \tilde{m}(t) \tag{5-89}$$

where A, G, and H are constant matrices of dimension $n \times n$, $n \times m$ ($m \leqq n$), and $r \times n$ ($r \leqq n$), respectively. The process noise $\tilde{n}(t)$ is a zero-mean Gaussian white-noise process with intensity for the covariance given by $Q = Q' \geqq 0$. The measurement noise $\tilde{m}(t)$ is also a zero-mean Gaussian white-noise process with intensity matrix $R = R' > 0$. The processes $\tilde{n}(t)$ and $\tilde{m}(t)$ along with the Gaussian $x(t_0)$ are mutually uncorrelated. The optimal filter problem is to construct an estimate $\hat{x}(t)$ of $x(t)$ operating over $(t_0,t]$ such that the quantity,

$$J = \lim_{t_0 \to -\infty} E\left[e'(t)\ W\ e(t)\right] \tag{5-90}$$

is minimized where E denotes expected value and

$$e(t) = x(t) - \hat{x}(t) \qquad (5\text{-}91)$$

The constant $n \times n$ matrix W satisfies

$$W = W' \overset{\geq}{=} 0 \qquad (5\text{-}92)$$

When the pair (A',H') is stabilizable and the (A',D') pair (where $DD' = GQG'$) is detectable, the solution to the asymptotic optimal observer problem exists and $\hat{x}(t)$ is given by

$$\dot{\hat{x}}(t) = A\,\hat{x}(t) + F\left[y(t) - H\,\hat{x}(t)\right] \qquad (5\text{-}93)$$

where

$$\hat{x}(t_0) = E\left[x(t_0)\right] \qquad (5\text{-}94)$$

with filter gain

$$F = PH'R^{-1} \qquad (5\text{-}95)$$

and P satisfying

$$0 = AP + PA' + GQG' - PH'R^{-1}HP \qquad (5\text{-}96)$$

The matrix $P = P' \overset{\geq}{=} 0$ represents the (constant) steady-state variance matrix of the reconstruction error $e(t)$; that is,

$$\lim_{t_0 \to -\infty} E\left[e(t)\,e'(t)\right] = P \qquad (5\text{-}97)$$

Also,

$$J = \lim_{t_0 \to -\infty} E\left[e'(t)\,W\,e(t)\right] = \text{trace }(PW) \qquad (5\text{-}98)$$

For the discrete case,

$$x(i+1) = A\ x(i) + G\ \tilde{n}(i) \qquad (5\text{-}99)$$

with output

$$y(i) = H\ x(i) + \tilde{m}(i) \qquad (5\text{-}100)$$

where A, G, and H are as previously defined. The process noise $\tilde{n}(i)$ $(i = i_0, i_0+1, \ldots)$ is a sequence of zero-mean Gaussian white-noise stochastic variables with variance matrix $Q = Q' \geq 0$. The measurement noise $\tilde{m}(i)$ $(i = i_0, i_0+1, \ldots)$ is also a zero-mean Gaussian white-noise [5-1] sequence with variance $R = R' > 0$. The processes $\tilde{n}(i)$ and $\tilde{m}(i)$ along with the Gaussian $x(i_0)$ are mutually uncorrelated. The optimal estimator problem considered here (technically a prediction problem) is to construct an estimate $\hat{x}(i)$ of $x(i)$ from knowledge of $y(i_0), y(i_0+1), \ldots, y(i-1)$ such that the quantity,

$$J = \lim_{i_0 \to -\infty} E\left[e'(i)\ W\ e(i)\right] \qquad (5\text{-}101)$$

is minimized, where

$$e(i) = x(i) - \hat{x}(i) \qquad (5\text{-}102)$$

with W as previously defined. When the (A', H') pair is stabilizable and the (A', D') pair (with $DD' = GQG'$) is detectable, the solution to the asymptotic optimal observer problem exists and $\hat{x}(i)$ is given by

$$\hat{x}(i+1) = A\ \hat{x}(i) + F\left[y(i) - H\ \hat{x}(i)\right] \qquad (5\text{-}103)$$

where

$$\hat{x}(i_0) = E\left[x(i_0)\right] \qquad (5\text{-}104)$$

with filter gain

$$F = APH'(R + HPH')^{-1} \qquad (5\text{-}105)$$

and P satisfying

$$P = \phi P\phi' + FRF' + GQG' \qquad (5\text{-}106)$$

for

$$\phi = A - FH \qquad (5\text{-}107)$$

The matrix $P = P' \geq 0$ represents the (constant) steady-state variance matrix of the reconstruction error $e(i)$; that is,

$$\lim_{i_0 \to -\infty} E\!\left[e(i)\ e'(i)\right] = P \qquad (5\text{-}108)$$

Also,

$$J = \lim_{i_0 \to -\infty} E\!\left[e'(i)\ W\ e(i)\right] = \text{trace } (PW) \qquad (5\text{-}109)$$

Computation for both the discrete and continuous versions of the foregoing optimal filter problems is performed using duality theory [5-1] and the regulator subroutine ASYMREG. Thus, the user has the option of solving the appropriate steady-state covariance equations in P by either of the subroutines DISCREG, CNTNREG, or RICTNWT. No computations are performed which involve the matrix W; hence, no data for W are required.

2. USAGE

a. Calling Sequence

CALL ASYMFIL(A,NA,G,NG,H,NH,Q,NQ,R,NR,F,NF,P,NP,IDENT,DISC,NEWT,STABLE,
FNULL,ALPHA,IOP,DUMMY)

b. Input Arguments

A, G, H, Q, R   Matrices packed by columns in one-dimensional arrays. Upon normal return, all except Q are not destroyed.

127

NA, NG, NH, NQ, NR	Two-dimensional vectors giving the number of rows and columns of respective matrices; for example,
	NA(1) = Number of rows of A
	NA(2) = Number of columns of A
	Not destroyed upon normal return except for NQ replaced by NA
F, NF, P, NP	F and P are matrices packed by columns into one-dimensional arrays of dimension at least rn and n^2, respectively.
	NF and NP are the corresponding two-dimensional vectors giving the number of rows and columns in F and P.
	Any input requirements depend on and are specified by whether DISCREG, CNTNREG, or RICTNWT is employed by ASYMREG. When F data are input, it should be kept in mind that, because of the use of duality theory, ASYMREG will treat A' as A and H' as B. An initial stabilizing gain F for use in RICTNWT must cause (A' - H'F) to be asymptotically stable and be appropriately dimensioned.
IDENT	Logical variable:
	TRUE If G is an identity matrix
	FALSE Otherwise
	If G is an identity matrix, no data need be input for G, but G and NG must still appear as arguments of the calling sequence.

DISC   Logical variable:

                TRUE   If the digital version is solved

                FALSE   For the continuous version

NEWT   Logical variable:

                TRUE   RICTNWT is used to solve the appropriate

                         steady-state covariance equation

                FALSE   Either CNTNREG or DISCREG is used depending

                         upon the value of DISC.

STABLE   Logical variable. Data for STABLE are not required if

                NEWT = FALSE, but STABLE must still appear as an argu-

                ment of the calling sequence. When NEWT = TRUE:

                STABLE = TRUE   If it is known that the matrix

                                  $(A' - H'F)$ computed from input

                                  data is stable relative to an input

                                  parameter ALPHA.

                STABLE = FALSE $(A' - H'F)$ is computed and tested

                                  (within ASYMREG) for stability

                                  relative to ALPHA.

FNULL   Logical variable. Data for FNULL are not required if

                NEWT = FALSE, but FNULL must still appear as an argument

                of the calling sequence. When NEWT = TRUE:

                FNULL = TRUE   If the input F is a null matrix

                FNULL = FALSE   Otherwise

ALPHA   Scalar variable. ALPHA is not required if NEWT = FALSE

                or STABLE = TRUE. Otherwise, ALPHA is used in the

                asymptotic stability test of $(A' - H'F)$ from input

                data.

IOP   Five-dimensional option vector:

                      IOP(1) = 0   Do not print results.

                      Otherwise   Print input, GQG', F, P, the eigenvalues of P, the matrix (A - FH), and the eigenvalues of (A - FH).

                      IOP(2), IOP(3), IOP(4), and IOP(5) are the first four elements of the IOP vector of ASYMREG.

DUMMY   Vector of working space for computations, dimensioned at least:

DISC	NEWT		
TRUE	TRUE	$6n^2 + 4n + 2$	(STABLE = FALSE)
		$5n^2$	(STABLE = TRUE)
TRUE	FALSE	$4n^2$	
FALSE	FALSE	$17n^2 + n(r + 18)$	
FALSE	TRUE	$8n^2 + n(r + 1)$	(BILIN)
		$5n^2 + n(r + 4) + 2$	(BARSTW)

c. Output Arguments

Q   Upon normal return, Q is replaced by GQG'. The dimension of Q must be at least n^2.

F   Upon normal return, F contains the filter gain for either the continuous or the discrete problem.

NF   Two-dimensional vector giving, upon normal return, the number of rows and columns of F:

                      NF(1) = n

                      NF(2) = r

P   Upon normal return, P contains the steady-state covariance matrix.

NP   Two-dimensional vector giving the number of rows and columns of P. Upon normal return, NP = NA.

DUMMY   Upon normal return, the first n elements of DUMMY contain the eigenvalues of P, the next n^2 contain the matrix $(A - FH)$ for filter gain F, and the next $2n$ contain the eigenvalues of $(A - FH)$ stored as an $n \times 2$ matrix with the real parts as first column and imaginary parts as the second. All matrices are packed by columns into one-dimensional arrays.

d. COMMON Blocks

None

e. Error Messages

None directly from ASYMFIL

f. Subroutines Employed by ASYMFIL

LNCNT, PRNT, TRANP, EQUATE, ASYMREG

g. Subroutines Employing ASYMFIL

None

h. Concluding Remarks

Cases in which G, with dimension $n \times m$, has $m \geq n$ can be treated as follows. Compute GQG' externally and, using subroutine FACTOR, find a $q \times n$ matrix D $(q \leq n)$ such that

$$GQG' = D'D \qquad (5\text{-}110)$$

Then apply ASYMFIL with G replaced by D' and $Q = I_n$.

Extensions of the basic optimal filter problem considered in ASYMFIL (such as colored noise, singular R matrices, and correlated process and measurement noise) can be found in the literature (e.g., [5-1]).

Generally, each extension can be solved by an appropriate combination of ASYMFIL with other subroutines of the ORACLS program. The transient Kalman-Bucy filter problem in which t_0 and i_0 are finite can be solved by using duality theory and the transient regulator solution capability of subroutines CNTNREG and DISCREG.

J/ Explicit Model Following (EXPMDFL)

Subroutine EXPMDFL solves either the continuous or discrete time-invariant asymptotic explicit (model-in-the-system) model-following problem [5-14].

For the continuous case, the state and output equations are given as

$$\dot{x}(t) = A\ x(t) + B\ u(t) \tag{5-111}$$

$$y(t) = H\ x(t) \tag{5-112}$$

where $x(0) = x_0$ is given and the constant matrices A, B, and H are of dimension $n \times n$, $n \times r$ $(r \leqq n)$, and $m \times n$ $(m \leqq n)$, respectively. The control function $u(t)$ is required to minimize

$$J = \lim_{t_1 \to \infty} \left\{ \int_0^{t_1} \left[e'(t)\ Q\ e(t) + u(t')\ R\ u(t) \right]\ dt \right\} \tag{5-113}$$

where

$$e(t) = y(t) - y_m(t) \tag{5-114}$$

$$y_m(t) = H_m\ x_m(t) \tag{5-115}$$

and

$$\dot{x}_m(t) = A_m\ x_m(t) \tag{5-116}$$

where $x_m(0) = x_m^o$ is given. The constant matrices H_m and A_m have

132

dimension $m \times \ell$ $(m \leq \ell)$ and $\ell \times \ell$, respectively. Also, $Q = Q' \geq 0$ and $R = R' > 0$. The optimization of the performance index causes the output $y(t)$ of the state to track the output $y_m(t)$ of a prescribed model. After substituting $e(t)$ into the performance index, the model-following problem can be transformed into choosing $u(t)$ to minimize

$$J = \lim_{t_1 \to \infty} \int_0^{t_1} \left[\tilde{x}'(t) \, \tilde{Q} \, \tilde{x}(t) + u'(t) \, R \, u(t) \right] \, dt \qquad (5\text{-}117)$$

with

$$\dot{\tilde{x}}(t) = \tilde{A} \, \tilde{x}(t) + \tilde{B} \, u(t) \qquad (5\text{-}118)$$

where

$$\tilde{A} = \begin{bmatrix} A & 0 \\ 0 & A_m \end{bmatrix} \qquad (5\text{-}119)$$

$$\tilde{B} = \begin{bmatrix} B \\ 0 \end{bmatrix} \qquad (5\text{-}120)$$

$$\tilde{Q} = \begin{bmatrix} H'QH & -H'QH_m \\ -H_m'QH & H_m'QH_m \end{bmatrix} \qquad (5\text{-}121)$$

and

$$\tilde{x} = \begin{bmatrix} x \\ x_m \end{bmatrix} \qquad (5\text{-}122)$$

This transformed problem can be solved directly using optimal linear regulator theory. If the $(\tilde{A},\tilde{B})$ pair is stabilizable and the $(\tilde{A},D)$ pair (with $D'D = \tilde{Q}$) is detectable, the solution exists and is given by

133

$$u(t) = -F\ \tilde{x}(t) = -F_{11}\ x(t) - F_{12}\ x_m(t) \tag{5-123}$$

Computationally, it is inefficient to work with the composite $(\tilde{A}, \tilde{B})$ system directly. If the steady-state Riccati equation is formed and $\tilde{A}$, $\tilde{B}$, and $\tilde{Q}$ are substituted, it readily follows [5-14] that

$$F_{11} = R^{-1}B'P_{11} \tag{5-124}$$

with $P_{11} = P_{11}' \geqq 0$ satisfying

$$P_{11}A + A'P_{11} - P_{11}BR^{-1}B'P_{11} + H'QH = 0 \tag{5-125}$$

and

$$F_{12} = R^{-1}B'P_{12} \tag{5-126}$$

with P_{12} satisfying

$$P_{12}A_m + (A - BF_{11})'P_{12} = H'QH_m \tag{5-127}$$

The computation of (F_{11}, F_{12}) thus separates into two parts:

(1) Evaluate the feedback gain F_{11} on the state x by solving a reduced-order optimal regulator problem of the form

$$\dot{x}(t) = A\ x(t) + B\ v(t) \tag{5-128}$$

$$y(t) = H\ x(t) \tag{5-112}$$

$$\min_{v(t)} \left\{ \lim_{t_1 \to \infty} \int_0^{t_1} \left[y'(t)\ Q\ y(t) + v'(t)\ R\ v'(t) \right]\ dt \right\} \tag{5-129}$$

leading to

$$v(t) = -F_{11}\ x(t) \tag{5-130}$$

(2) Using F_{11} from step (1), compute the feedforward gain F_{12} on the model x_m from the linear equations

$$P_{12}A_m + (A - BF_{11})'P_{12} = H'QH_m \qquad (5\text{-}127)$$

$$RF_{12} = B'P_{12} \qquad (5\text{-}131)$$

For the discrete case, the state and output equations are given as

$$x(i+1) = A\ x(i) + B\ u(i) \qquad (5\text{-}132)$$

$$y(i) = H\ x(i) \qquad (5\text{-}133)$$

with A, B, and H as previously defined. The control sequence $u(i)$ $(i = 0,1,\ldots,N\text{-}1)$ is required to minimize

$$J = \lim_{N\to\infty}\left\{\sum_{i=0}^{N-1}\Big[e'(i+1)\ Q\ e(i+1) + u'(i)\ R\ u(i)\Big]\right\} \qquad (5\text{-}134)$$

where

$$e(i) = y(i) - y_m(i) \qquad (5\text{-}135)$$

$$y_m(i) = H_m\ x_m(i) \qquad (5\text{-}136)$$

$$x_m(i+1) = A_m\ x_m(i) \qquad (5\text{-}137)$$

with Q, R, H_m, and A_m as previously defined. As in the continuous case, the discrete model-following problem can be solved in terms of a $(\tilde{A},\tilde{B},\tilde{Q},\tilde{R})$ optimal regulator formulation, but a simplified computational algorithm also exists [5-15]:

(1) Compute a feedback gain F_{11} on the state x by solving the reduced-order optimal regulator problem

$$x(i+1) = A\ x(i) + B\ v(i) \qquad (5\text{-}138)$$

$$y(i) = H\ x(i) \qquad (5\text{-}133)$$

$$\min_{v(i)} \left\{ \lim_{N \to \infty} \sum_{i=0}^{N-1} \left[y'(i+1) \ Q \ y(i+1) + v'(i) \ R \ v(i) \right] \right\} \qquad (5\text{-}139)$$

leading to

$$v(i) = -F_{11} \ x(i) \qquad (5\text{-}140)$$

(2) Using (P_{11}, F_{11}) from step (1), compute a feedforward gain F_{12} on the model state x_m from the linear equations,

$$P_{12} = (A - BF_{11})' P_{12} A_m - H'QH_m \qquad (5\text{-}141)$$

$$(B'P_{11}B + R)F_{12} = B'P_{12}A_m \qquad (5\text{-}142)$$

The complete optimal model-following control law is then given by

$$u(i) = -F_{11} \ x(i) - F_{12} \ x_m(i) \qquad (5\text{-}143)$$

EXPMDFL solves both the continuous and discrete versions of the explicit model-following problem through the simplified computational approach of solving a reduced-order regulator program for F_{11} and a set of linear equations for F_{12}. Subroutine ASYMREG is used to solve the reduced-order regulator problem, thus giving the user the choice of solving the appropriate steady-state Riccati equation for P_{11} by either of the subroutines DISCREG, CNTNREG, or RICTNWT. The subroutine BARSTW is used to solve the P_{12} equation in the continuous case and subroutine SUM in the discrete case. Final F_{12} computation in both cases employs subroutine SYMPDS. Computational parameters for BARSTW are set internally through the COMMON block TOL of subroutine RDTITL. An option is provided to bypass the (P_{11}, F_{11}) computation and, using input (P_{11}, F_{11}) data, proceed directly to the computation of (P_{12}, F_{12}).

136

2. USAGE

a. Calling Sequence

CALL EXPMDFL(A,NA,B,NB,H,NH,AM,NAM,HM,NHM,Q,NQ,R,NR,F,NF,P,NP,HIDENT,

HMIDENT,DISC,NEWT,STABLE,FNULL,ALPHA,IOP,DUMMY)

b. Input Arguments

A, B, H, AM, HM, Q, R	Matrices packed by columns in one-dimensional arrays; not destroyed upon normal return
NA, NB, NH, NAM, NHM, NQ, NR	Two-dimensional vectors giving the number of rows and columns of the respective matrices; for example, $NA(1)$ = Number of rows of A $NA(2)$ = Number of columns of A Not destroyed upon normal return
F	Matrix packed by columns into an array dimensioned at least $r(n + \ell)$. The first rn elements of F are treated as an input matrix F to ASYMREG for solving the reduced-order Riccati equation in P_{11}. If no input F data are required by ASYMREG, no data need be input for F in EXPMDFL unless the reduced-order Riccati computation for (P_{11}, F_{11}) is to be bypassed. In this case, the first rn elements of F should contain a matrix F_{11} to be used in the computation of $(A - BF_{11})$.
NF	Two-dimensional vector giving, if required, the number of rows and columns of F used by ASYMREG or F_{11}: $NF(1) = r$ $NF(2) = n$
P	Matrix packed by columns into an array dimensioned at least $n(n + \ell)$. The first n^2 elements of P are

treated as an input matrix P to ASYMREG when used to solve the reduced-order Riccati equation for P_{11}. If no input P data are required by ASYMREG, no data need be input for P in EXPMDFL unless the reduced-order Riccati computation for (P_{11}, F_{11}) is to be bypassed in the discrete case. In the discrete case, the first n^2 elements must contain a matrix to be used for P_{11} in the computation of the discrete F_{12}.

NP   Two-dimensional vector giving, if required, the number of rows and columns of P used by ASYMREG or P_{11}:

$NP(1) = n$

$NP(2) = n$

HIDENT   Logical variable:

TRUE   If H is an identity matrix

FALSE   Otherwise

If H is an identity matrix, no data need be input for H, but H and NH must still appear as arguments of the calling sequence.

HMIDENT   Logical variable:

TRUE   If H_m (HM) is an identity matrix

FALSE   Otherwise

If H_m is an identity matrix, no data need be input for HM, but HM and NHM must still appear as arguments of the calling sequence.

DISC   Logical variable:

TRUE   If the discrete version is solved

FALSE   For the continuous version

NEWT, STABLE, FNULL, ALPHA	Variables whose input values are determined by the choice of method to solve the steady-state Riccati equation for (P_{11}, F_{11}) called for in ASYMREG; not required if the (P_{11}, F_{11}) computation is bypassed but still must appear as arguments of the calling sequence

IOP   Five-dimensional option vector:

         IOP(1) = 0   Do not print results.

         Otherwise   Print input and computed results.

         IOP(2) = 0   Do not compute (P_{11}, F_{11}) but use input P and F as these variables.

         Otherwise   Compute P_{11} and F_{11} through ASYMREG.

         IOP(3), IOP(4), and IOP(5) are first three elements of the IOP vector in ASYMREG; not required if IOP(2) = 0.

DUMMY   Vector of working space for computations, dimensioned at least:

IOP(2)	DISC	
0	TRUE	$3n^2 + \ell^2 + \max(n^2, \ell n)$
0	FALSE	$3n^2 + 4n + 2 + \max(n^2, \ell n)$
Nonzero	TRUE or FALSE	Either n^2 plus the DUMMY requirement of ASYMREG or the preceding requirements for IOP(2) = 0, whichever is larger

c. Output Arguments

F   Upon normal return, $F = \begin{bmatrix} F_{11}, F_{12} \end{bmatrix}$ packed by columns in a one-dimensional array

NF	Upon normal return,

$$NF(1) = r$$

$$NF(2) = n + \ell$$

P	For $IOP(2) \neq 0$, $P = \begin{bmatrix} P_{11}, P_{12} \end{bmatrix}$ upon normal return.

If $IOP(2) = 0$, the first $n\ell$ elements contain the matrix P_{12}. In both cases, P is packed by columns into a one-dimensional array.

NP	Upon normal return,

$$NP(1) = n$$

$$NP(2) = n + \ell \qquad (IOP(2) \neq 0)$$

or

$$NP(2) = \ell \qquad (IOP(2) = 0)$$

d. COMMON Block

TOL

e. Error Message

If either R or $(B'P_{11}B+R)$ fails to be positive definite, the message "IN EXPMDFL, THE COEFFICIENT MATRIX FOR SYMPDS IS NOT SYMMETRIC POSITIVE DEFINITE" is printed, and the program is returned to the calling point.

f. Subroutines Employed by EXPMDFL

LNCNT, PRNT, EQUATE, ASYMREG, MULT, SUBT, TRANP, BARSTW, SUM, ADD, SYMPDS, SCALE

g. Subroutines Employing EMPMDFL

None

h. Concluding Remarks

If the model dynamics in both the continuous and discrete cases contain a direct link to the control, then

$$\tilde{B} = \begin{bmatrix} B \\ B_m \end{bmatrix} \qquad (5\text{-}144)$$

with $B_m \neq 0$, and the computation will not generally decouple into a simplified algorithm. In this case, ASYMREG can be applied to the composite $(\tilde{A}, \tilde{B}, \tilde{Q}, R)$ system directly. For the transient case where N and t_1 are finite, the time-varying explicit model-following solution can be obtained from ORACLS by using the composite $(\tilde{A}, \tilde{B}, \tilde{Q}, R)$ system and subroutine DISCREG or CNTNREG.

K/ Implicit Model Following (IMPMDFL)

1. PURPOSE

Subroutine IMPMDFL solves either the continuous or discrete time-invariant asymptotic implicit (model-in-the-performance-index) model-following problem.

For the continuous case, the state and output equations are given as

$$\dot{x}(t) = A x(t) + B u(t) \qquad (5\text{-}145)$$

$$y(t) = H x(t) \qquad (5\text{-}146)$$

where $x(0) = x_0$ is given and the constant matrices A, B, and H are of dimension $n \times n$, $n \times r$ $(r \leq n)$, and $m \times n$ $(m \leq n)$, respectively. The control function $u(t)$ is required to minimize

$$J = \lim_{t_1 \to \infty} \left\{ \int_0^{t_1} \left[e'(t) Q e(t) + u'(t) R u(t) \right] dt \right\} \qquad (5\text{-}147)$$

where

$$e(t) = \dot{y}(t) - A_m y(t) - B_m u(t) \qquad (5\text{-}148)$$

141

The constant matrices A_m and B_m have dimension $m \times m$ and $m \times r$, respectively. Also, $Q = Q' \overset{>}{=} 0$ and $R = R' > 0$. The philosophy of implicit model following is to cause the output $y(t)$ of the system state to behave similar to a model state with dynamics

$$\dot{x}_m(t) = A_m x_m(t) + B_m u(t) \tag{5-149}$$

where $x_m(0) = x_m^o$ is given. This is done by forcing, in a weighted least squares sense, $y(t)$ to satisfy the model dynamic equation. Substituting for $e(t)$ in the performance index reduces the problem to choosing $u(t)$ to minimize

$$J = \lim_{t_1 \to \infty} \left\{ \int_0^{t_1} \left[x'(t) \, \tilde{Q} \, x(t) + x'(t) \, \tilde{W} \, u(t) + u'(t) \, \tilde{R} \, u(t) \right] dt \right\} \tag{5-150}$$

where

$$\tilde{Q} = (HA - A_m H)'Q(HA - A_m H) \tag{5-151}$$

$$\tilde{W} = 2(HA - A_m H)'Q(HB - B_m) \tag{5-152}$$

and

$$\tilde{R} = (HB - B_m)'Q(HB - B_m) + R \tag{5-153}$$

Applying the control transformation

$$u(t) = -F_0 \, x(t) + v(t) \tag{5-154}$$

with

$$F_0 = \tilde{R}^{-1} \frac{\tilde{W}'}{2} \tag{5-155}$$

further reduces the problem to choosing $v(t)$ to minimize

$$J = \lim_{t_1 \to \infty} \left\{ \int_0^{t_1} \left[x'(t) \, \hat{Q} \, x(t) + v'(t) \, \tilde{R} \, v(t) \right] dt \right\} \tag{5-156}$$

where

$$\hat{Q} = \tilde{Q} - \frac{\tilde{W}}{2} F_0 \tag{5-157}$$

$$\tilde{A} = A - BF_0 \tag{5-158}$$

and

$$\dot{x}(t) = \tilde{A} \, x(t) + B \, v(t) \tag{5-159}$$

If the (A,B) pair is stabilizable and the $(\tilde{A},D)$ pair (with $D'D = \hat{Q}$) is detectable, a solution to the asymptotic implicit model-following problem exists and is given by

$$v(t) = -F_1 \, x(t) \tag{5-160}$$

$$F_1 = \tilde{R}^{-1}B'P \tag{5-161}$$

$$P\tilde{A} + \tilde{A}'P + Q - PB\tilde{R}^{-1}B'P = 0 \tag{5-162}$$

and

$$u(t) = -F \, x(t) \tag{5-163}$$

with

$$F = F_0 + F_1 \tag{5-164}$$

For the discrete case, the state and output equations are given as

$$x(i+1) = A \, x(i) + B \, u(i) \tag{5-165}$$

$$y(i) = H \, x(i) \tag{5-166}$$

where $x(0) = x_0$ is given and A, B, and H are as previously defined. The control function $u(i)$ $(i = 0,1,\ldots,N-1)$ is required to minimize

$$J = \lim_{N \to \infty} \left\{ \sum_{i=0}^{N-1} \left[e'(i) \, Q \, e(i) + u'(i) \, R \, u(i) \right] \right\} \tag{5-167}$$

143

where

$$e(i) = y(i+1) - A_m y(i) - B_m u(i) \qquad (5-168)$$

The matrices A_m, B_m, Q, and R are as previously defined. Substituting for $e(i)$ in the performance index reduces the problem to choosing $u(i)$ to minimize

$$J = \lim_{N \to \infty} \left\{ \sum_{i=0}^{N-1} \left[x'(i) \, \tilde{Q} \, x(i) + x'(i) \, \tilde{W} \, u(i) + u'(i) \, \tilde{R} \, u(i) \right] \right\} \qquad (5-169)$$

Applying the control transformation,

$$u(i) = -F_0 \, x(i) + v(i) \qquad (5-170)$$

further reduces the problem to choosing $v(i)$ to minimize

$$J = x'(0) \, \hat{Q} \, x(0) + \lim_{N \to \infty} \left\{ \sum_{i=0}^{N-1} \left[x'(i+1) \, \hat{Q} \, x(i+1) + v'(i) \, \hat{R} \, v(i) \right] \right\} \qquad (5-171)$$

with

$$x(i+1) = \tilde{A} \, x(i) + B \, v(i) \qquad (5-172)$$

If the aforementioned stabilizability and detectability conditions are satisfied

$$v(i) = -F_1 \, x(i) \qquad (5-173)$$

$$F_1 = (\tilde{R} + B'PB)^{-1} B'P \, \tilde{A} \qquad (5-174)$$

$$P = \phi'P\phi + F_1' \, \tilde{R} \, F_1 + \hat{Q} \qquad (5-175)$$

$$\phi = \tilde{A} - BF_1 \qquad (5-176)$$

and

$$u(i) = -F \, x(i) \qquad (5-177)$$

with

$$F = F_0 + F_1 \qquad\qquad (5\text{-}178)$$

The program IMPMDFL solves the steady-state Riccati equation to generate the appropriate P and F_1 through use of the subroutine ASYMREG, thus giving the user the choice of either subroutine DISCREG, CNTNREG, and RICTNWT to solve the final Riccati equation. An option is provided to bypass the (P,F_1) computation and return after computing $\tilde{Q}$, $\tilde{W}$, $\tilde{R}$, F_0, $\hat{Q}$, and $\tilde{A}$.

2. USAGE

a. Calling Sequence

CALL IMPMDFL(A,NA,B,NB,H,NH,AM,NAM,BM,NBM,Q,NQ,R,NR,F,NF,P,NP,IDENT,DISC, NEWT,STABLE,FNULL,ALPHA,IOP,DUMMY)

b. Input Arguments

A, B, H, AM, Matrices packed by columns in one-dimensional arrays;
BM, Q, R
       not destroyed upon normal return

NA, NB, NH, Two-dimensional vectors giving the number of rows and
NAM, NBM,
NQ, NR   columns of respective matrices; for example,

    NA(1) = Number of rows of A

    NA(2) = Number of columns of A

    Not destroyed upon normal return

F, NF, P, NP F and P are matrices packed by columns into one-
    dimensional arrays of dimension at least rn and n^2,
    respectively.

    NF and NP are the corresponding two-dimensional
    vectors giving the number of rows and columns of F
    and P.

Any input requirements depend on and are specified by
whether DISCREG, CNTNREG, or RICTNWT is employed by
ASYMREG. When F data are input for RICTNWT, it should
be such that the input $(\tilde{A},B)$ system is stabilized.

IDENT   Logical variable:

TRUE   If H is an identity matrix

FALSE   Otherwise

If H is an identity matrix, no data need be input
for H, but H and NH must still appear as arguments of the calling sequence.

DISC   Logical variable:

TRUE   If the discrete version is solved

FALSE   For the continuous version

NEWT, STABLE,   Variables whose input values are determined by the
FNULL, ALPHA
                  choice of method called for from ASYMREG to solve the
steady-state Riccati equation for P and F

IOP   Four-dimensional option vector:

$IOP(1) = 0$ Do not print results.

Otherwise   Print input and computed results.

$IOP(2)$, $IOP(3)$, and $IOP(4)$ are the first three elements of the IOP vector in ASYMREG. If $IOP(2) = -1000$,
the IMPMDFL program returns just prior to employing
ASYMREG in which case data for F, NF, P, NP, DISC,
NEWT, STABLE, FNULL, ALPHA, $IOP(3)$, and $IOP(4)$ need
not be entered, but these arguments must still appear
in the calling sequence.

DUMMY   Vector of working space for computations, dimensioned at
                 least $4n^2$ plus the dimension requirements for DUMMY
                 in ASYMREG when (P,F_1) is to be computed. If the
                 (P,F_1) computation is bypassed, DUMMY should be
                 dimensioned at least $6n^2 + r$.

c. Output Arguments

F   If (P,F_1) is computed, $F = F_0 + F_1$ upon normal
                 return. For (P,F_1) not computed, the input F is
                 returned. All matrices are packed by columns into a
                 one-dimensional array.

NF   Upon normal return,

                 $NF(1) = r$

                 $NF(2) = n$

P   If (P,F_1) is computed, P contains, upon normal
                 return, the solution to the steady-state Riccati
                 equation defining F_1. For (P,F_1) not computed, the
                 input P is returned. All matrices are packed by
                 columns into one-dimensional arrays.

NP   Upon normal return,

                 $NP(1) = n$

                 $NP(2) = n$

                 when (P,F_1) is computed. Otherwise, the input NP
                 is returned.

DUMMY   Upon normal return, if $IOP(2) = -1000$, the first n^2
                 elements contain the matrix $\hat{Q}$ and the next nr con-
                 tain the matrix F_0. After $2n^2$ elements, the next
                 r^2 contain the matrix $\tilde{R}$. After $3n^2$, the next n^2

contain the matrix $\tilde{A}$. Additionally, for (P, F_{11}) computed, after $4n^2$, the elements of DUMMY in IMPMDFL contain the elements of DUMMY returned by ASYMREG. All matrices are packed by columns as one-dimensional arrays.

d. COMMON Blocks

None

e. Error Messages

None directly from IMPMDFL

f. Subroutines Employed by IMPMDFL

LNCNT, PRNT, SUBT, EQUATE, PREFIL, ASYMREG, ADD, MULT, TRANP, SCALE

g. Subroutines Employing IMPMDFL

None

h. Concluding Remarks

For the case in which t_1 or N is finite, set $IOP(2) = -1000$ and return with $\hat{Q}$, F_0, $\tilde{R}$, and $\tilde{A}$ in DUMMY. Then, using the $(\tilde{A}, B, \hat{Q}, \tilde{R})$ system, solve for the transient $P(\ )$ and $F_1(\ )$ from subroutine DISCREG or CNTNREG. The model-following control law is then

$$F(\ ) = F_0 + F_1(\ ) \tag{5-179}$$

L/ Eigenvalue Placement (POLE)

1. PURPOSE

Given real matrices A and B of dimension $(n \times n)$ and $(n \times 1)$, respectively, subroutine POLE calculates an $(1 \times n)$ matrix F which causes the eigenvalues of the matrix $A-BF$ to assume n specified values. The controllability matrix

$$C = \begin{bmatrix} B, & AB, & \ldots, & A^{n-1}B \end{bmatrix} \tag{5-180}$$

148

is assumed to be nonsingular. Denote the prescribed eigenvalues of $A-BF$ by λ_i ($i = 1, \ldots, n$), with the restriction that complex values occur in conjugate pairs which are grouped consecutively in the λ_i sequence, and define coefficients $\tilde{\alpha}_i$ ($i = 1, \ldots, n$), through

$$(s - \lambda_1)(s - \lambda_2) \ldots (s - \lambda_n) = s^n + \tilde{\alpha}_1 s^{n-1} + \ldots + \tilde{\alpha}_{n-1} s + \tilde{\alpha}_n \qquad (5\text{-}181)$$

Also, denote the characteristic polynomial of A by

$$\text{Det. } (sI - A) = s^n + \alpha_1 s^{n-1} + \ldots + \alpha_{n-1} s + \alpha_n \qquad (5\text{-}182)$$

Then, it follows from [5-1] that F is obtained by solving

$$FV = \tilde{\alpha}' - \alpha' \qquad (5\text{-}183)$$

where

$$\tilde{\alpha}' = (\tilde{\alpha}_n, \tilde{\alpha}_{n-1}, \ldots, \tilde{\alpha}_1) \qquad (5\text{-}184)$$

$$\alpha' = (\alpha_n, \alpha_{n-1}, \ldots, \alpha_1) \qquad (5\text{-}185)$$

and

$$V = C \begin{bmatrix} \alpha_{n-1} & \alpha_{n-2} & \cdots & \alpha_1 & 1 \\ \alpha_{n-2} & \alpha_{n-1} & \cdots & 1 & 0 \\ \cdot & \cdot & & \cdot & \cdot \\ \cdot & \cdot & & \cdot & \cdot \\ \cdot & \cdot & & \cdot & \cdot \\ \alpha_1 & 1 & \cdots & 0 & 0 \\ 1 & 0 & \cdots & 0 & 0 \end{bmatrix} \qquad (5\text{-}186)$$

The computational procedure employed by subroutine POLE is:

(1) Form the controllability matrix C given by (5-180) and also obtain the product $A^n B$.

(2) Evaluate the coefficients α_i (i = 1, ..., n), of the characteristic polynomial of A. The Cayley-Hamilton Theorem [5-16] states that

$$A^n + \alpha_1 A^{n-1} + \alpha_{n-1} A + \alpha_n I = 0 \qquad (5\text{-}187)$$

where I and 0 are $n \times n$ identity and null matrices, respectively. Post multiplying equation (5-187) by B and rearranging gives

$$C\alpha = -A^n B \qquad (5\text{-}188)$$

which is used to determine α in (5-183).

(3) Form V from equation (5-186).

(4) The coefficients $\tilde{\alpha}_i$ (i = 1, ..., n), are calculated using Newton Sums and Identities [5-17]. Newton Sums, S_i (i = 1, ..., n), are defined by

$$S_i = \sum_{j=1}^{n} (\lambda_j)^i = \lambda_1^i + \lambda_2^i + \ldots + \lambda_n^i \qquad (5\text{-}189)$$

and the coefficients $\tilde{\alpha}_i$ are generated sequentially by the Newton Identities

$$\tilde{\alpha}_i = -\frac{1}{i}\left[S_i + \tilde{\alpha}_1 S_{i-1} + \ldots + \tilde{\alpha}_{i-1} S_1\right] \quad (i = 1,\ldots,n) \qquad (5\text{-}190)$$

Complex arithmetic in (5-189) is avoided for conjugate pairs $(\lambda_i, \lambda_{i+1}) = (\lambda_i, \bar{\lambda}_i)$ by noting that

$$\lambda_i^k + \lambda_{i+1}^k = \lambda_i^k + \bar{\lambda}_i^k = \text{TRACE}\left\{\begin{bmatrix} \text{Re}(\lambda_i) & \text{Im}(\lambda_i) \\ -\text{Im}(\lambda_i) & \text{Re}(\lambda_i) \end{bmatrix}^k\right\} \qquad (5\text{-}191)$$

(5) Finally, the matrix F is obtained as the transpose of the solution, X, of

150

$$V'X = \tilde{\alpha} - \alpha \qquad (5\text{-}192)$$

2. USAGE

a. Calling Sequence

CALL POLE(A,NA,B,NB,EVAL,NUMR,F,NF,IOP,DUMMY)

b. Input Arguments

A, B   Matrices packed by columns in one-dimensional arrays of dimension n^2 and n, respectively, where n is the order of the A matrix; not destroyed upon normal return

NA, NB   Two-dimensional vectors giving the number of rows and columns of respective matrices; for example,

     NA(1) = Number of rows of A

     NA(2) = Number of columns of A

     Not destroyed upon normal return

EVAL   Real one-dimensional array of dimensional at least n containing the prescribed eigenvalues of A with the real eigenvalues placed first followed by the real and imaginary parts of the complex eigenvalues placed consecutively as 2×1 subvectors.

NUMR   Number of real prescribed eigenvalues placed first in EVAL. Set NUMR = 0 if all eigenvalues are to be complex.

IOP   Print option parameter:

     IOP = 0   No printed output

     Otherwise   Print input and computed results

DUMMY   Vector of working space for computations, dimensioned at least $5n^2 + 7n$

c. Output Arguments

F   One-dimensional array of dimension at least n. Upon normal return, F contains the vector $V^{-1}(\tilde{\alpha}' - \alpha')$

151

NF   Two-dimensional vector holding, upon normal return, the number

          of rows and columns in F:

          NF(1) = 1

          NF(2) = n

DUMMY   Upon normal return, the first n^2 elements contain the matrix

          A-BF, the next 2n the computed eigenvalues of A-BF stored

          as an n × 2 matrix with first column the real parts and

          imaginary parts in the second, and the next n^2 the eigen-

          vectors of A-BF stored as the n × n real matrix output

          from subroutine EIGEN. All matrices are packed by columns as

          one-dimensional arrays.

d. COMMON Blocks

None

e. Error Messages

(1) If the computation of α in (5-188) fails, the message "IN POLE,

     THE CONTROLLABILITY MATRIX, C, HAS BEEN DECLARED SINGULAR BY

     GELIM" is printed and the program is returned to the calling point.

(2) If the computation of X in (5-192) fails, the message "IN POLE,

     THE MATRIX V HAS BEEN DECLARED SINGULAR BY GELIM" is printed and

     the program is returned to the calling point.

(3) If the eigensystem computation for A-BF fails, the message "IN

     POLE, AFTER CALL TO EIGEN WITH A-BF, IERR = _____" is printed

     and the calculations continue.

f. Subroutines Employed by POLE

LNCNT, PRNT, EQUATE, MULT, JUXTC, SCALE, GELIM, NULL, TRCE, SUBT,

TRANP, EIGEN

g. Subroutines Employing POLE

None

h. Concluding Remarks

None

REFERENCES

5-1. H. Kwakernaak and R. Sivan, Linear Optimal Control Systems, John Wiley and Sons, Inc., c.1972.

5-2. P. Dorato and A. H. Levis, "Optimal Linear Regulators: The Discrete-Time Case," IEEE Trans. Autom. Control, vol. AC-16, no. 6, Dec. 1971, pp. 613-620.

5-3. A. H. Levis, R. A. Schlueter, and M. Athans, "On the Behavior of Optimal Linear Sampled-Data Regulators," Int. J. Control, vol. 13, no. 2, Feb. 1971, pp. 343-361.

5-4. E. S. Armstrong, "Series Representations for the Weighting Matrices in the Sampled-Data Optimal Linear Regulator Problem," IEEE Trans. Autom. Control, vol. AC-23, no. 3, June 1978, pp. 478-479.

5-5. D. Kleinman, "On an Iterative Technique for Riccati Equations," IEEE Trans. Autom. Control, vol. AC-13, no. 1, Feb. 1968, pp. 114-115.

5-6. E. S. Armstrong and G. T. Rublein, "A Discrete Analog of the Extended Bass Algorithm for Stabilizing Constant Linear Systems," Proceedings of the 1976 IEEE Conference on Decision and and Control, 76CH1150-2CS, Dec. 1976, pp. 1129-1131.

5-7. E. S. Armstrong, "An Extension of Bass' Algorithm for Stabilizing Linear Continuous Constant Systems," IEEE Trans. Autom. Control, vol. AC-21, no. 4, Aug. 1976, pp. 629-631.

5-8. G. A. Hewer, "An Iterative Technique for the Computation of the Steady State Gains for the Discrete Optimal Regulator," IEEE Trans. Autom. Control, vol. AC-16, no. 4, Aug. 1971, pp. 382-383.

5-9. E. S. Armstrong and G. T. Rublein, "A Stabilization Algorithm for Linear Discrete Constant Systems," IEEE Trans. Autom. Control, vol. AC-20, no. 1, Feb. 1975, pp. 153-154.

5-10. K. J. Åström, Introduction to Stochastic Control, Academic Press, 1970.

5-11. D. R. Vaughn, "A Negative Exponential Solution for the Matrix Riccati Equation," IEEE Trans. Autom. Control, vol. AC-14, no. 1, Feb. 1969, pp. 72-75.

5-12. L. V. Kantorovich and G. P. Akilov (D. E. Brown, trans.), Functional Analysis in Normed Spaces, (A. P. Robertson, ed.), MacMillian Co., 1964.

5-13. N. R. Sandell, Jr., "On Newton's Method for Riccati Equation Solution," IEEE Trans. Autom. Control, vol. AC-19, no. 3, June 1974, pp. 254-255.

5-14. B. D. O. Anderson and J. B. Moore, Linear Optimal Control, Prentice Hall, Inc., c.1971.

5-15. E. S. Armstrong, "Digital Explicit Model Following With Unstable Model Dynamics," AIAA Paper No. 74-888, AIAA Mechanics and Control of Flight Conference, Aug. 1974.

5-16. P. Lancaster, Theory of Matrices, Academic Press, c.1969.

5-17. J. V. Uspensky, Theory of Equations, McGraw-Hill Book Company, Inc., c.1948.

SIX/ SUPPORTING PROGRAMS

I/ INTRODUCTION

This chapter describes ORACLS subroutines that are used primarily to support the procedures described in previous chapters. Subroutine READ1 is an author-modified version of similarly titled software from the VASP program [6-1]. Subroutines AXPXB, SHRSLV, ATXPXA, SYMSLV, HSHLDR, BCKMLT, SCHUR, and SYSSLV were coded directly from the Bartels and Stewart article [6-2]. Software for the remaining subroutines was obtained from the Analysis and Computation Division subprogram library at the Langley Research Center.

II/ SUBROUTINE DESCRIPTIONS
A/ Input Numerical Data (READ1)

1. PURPOSE

Subroutine READ1 inputs from cards and prints or prints without card input a single matrix A without employing a header card as in subroutine READ. When card input is employed, each row of the matrix starts on a new card using format (8F10.2). Printing is performed using subroutine PRNT.

2. USAGE

a. Calling Sequence

CALL READ1(A,NA,NZ,NAM)

b. Input Arguments

A   Matrix packed by columns in a one-dimensional array; required
          as input only if $NZ(1) \neq 0$

NA, NZ   Two-dimensional arrays generally giving the number of rows and
          columns of A:

          $NA(1) = NZ(1)$ = Number of rows of A

          $NA(2) = NZ(2)$ = Number of columns of A

          Alternatively, if $NZ(1) = 0$, the data for (A,NA) are

          assumed to have been previously stored in locations A and

          NA. Otherwise, data for NA are not required as input.

NAM   Hollerith data: a four-character symbol used by PRNT

c. Output Arguments

A, NA   For $NZ(1) \neq 0$, the matrix A which was read from cards and
          packed by columns into a one-dimensional array; upon normal
          return,

          $NA(1) = NZ(1)$

          $NA(2) = NZ(2)$

d. COMMON Blocks

None

e. Error Message

If either $NZ(1)$ or $NZ(1) \times NZ(2)$ is less than unity, the message
"ERROR IN READ1 MATRIX _____ HAS NA = _____, _____" is printed, and
the program is returned to the calling point.

f. Subroutines Employed by READ1

LNCNT, PRNT

g. Subroutine Employing READ1

READ

h. Concluding Remarks

None

B/ Balance Square Matrix (BALANC)

1. PURPOSE

Subroutine BALANC balances a real square matrix for the calculation of
eigenvalues and eigenvectors and isolates the eigenvalues whenever
possible. The computational method follows that of Parlett and Reinsch
[6-3]. The subroutine BALANC is a translation of the ALGOL procedure
BALANCE found on pages 315 to 326 of [6-4].

2. USAGE

a. Calling Sequence

CALL BALANC(NM,N,A,LOW,IGH,SCALE)

b. Input Arguments

NM   First dimension of array A as given in the calling program

N   Number of rows of matrix A

A   Matrix which is to be balanced stored in a real two-dimensional
            array. The contents of this array are destroyed upon return.

c. Output Arguments

A   Upon normal return, A contains the balanced matrix.

LOW, IGH   Two integers such that $A(I,J) = 0$ if $I > J$ and
            $J = 1,2,\ldots,LOW-1$ or $I = IGH+1,IGH+2,\ldots,N$.

SCALE   A one-dimensional array of dimension at least N. SCALE con-
            tains information determining the permutations and scaling
            factors used: suppose that the principal submatrix in rows
            LOW through IGH has been balanced, that $P(J)$ denotes the
            index interchanged with J during the permutation step,
            and that the elements of the diagonal matrix used are

denoted by $D(I,J)$; then,

$$SCALE(J) = P(J) \qquad (J = 1,2,\ldots,LOW-1)$$
$$= D(J,J) \qquad (J = LOW,LOW+1,\ldots,IGH)$$
$$= P(J) \qquad (J = IGH+1,IGH+2,\ldots,N)$$

The order in which the interchanges are made is N to $IGH + 1$, then 1 to $LOW-1$.

d. COMMON Blocks

None

e. Error Messages

None directly from BALANC

f. Subroutines Employed by BALANC

None

g. Subroutines Employing BALANC

EIGEN

h. Concluding Remarks

None

C/ Upper Hessenberg Form (ELMHES)

1. PURPOSE

Given a real square matrix A, subroutine ELMHES reduces a submatrix situated in rows and columns LOW through IGH to upper Hessenberg form by stabilized elementary similarity transformations. The computational method follows that of Martin and Wilkinson [6-5]. ELMHES is a translation of the ALGOL procedure ELMHES found on pages 339 to 358 of [6-4].

2. USAGE

a. Calling Sequence

CALL ELMHES(NM,N,LOW,IGH,A,INT)

b. Input Arguments

NM   First dimension of the array A as given in the calling program

N   Number of rows of matrix A

LOW, IGH   Integers typically determined by the balancing subroutine BALANC for eigenvalue and eigenvector computation. If BALANC is not used, set LOW = 1 and IGH = N.

A   Matrix which is to be used in the reduction to upper Hessenberg form stored in a real two-dimensional array. The contents of A are destroyed upon return.

c. Output Arguments

A   Upon normal return, A contains the upper Hessenberg matrix. The multipliers which were used in the reduction are stored in the remaining triangle under the Hessenberg matrix.

INT   One-dimensional integer array of dimension at least IGH in the calling program. INT contains information on the rows and columns interchanged in the reduction. Only elements LOW to IGH are used.

d. COMMON Blocks

None

e. Error Messages

None directly from ELMHES

f. Subroutines Employed by ELMHES

None

g. Subroutine Employing ELMHES

EIGEN

h. Concluding Remarks

None

D/ Eigenvalues (HQR)

1. PURPOSE

Subroutine HQR finds the eigenvalues of a real square upper Hessenberg matrix H by the QR algorithm. The computational method follows that of Martin, Peters, and Wilkinson [6-6]. The subroutine HQR is a translation of the ALGOL procedure HQR found on pages 359 to 371 of [6-4].

2. USAGE

a. Calling Sequence

CALL HQR(NM,N,LOW,IGH,H,WR,WI,IERR)

b. Input Arguments

NM   First dimension of array H as given in the calling
            program

N   Number of rows in matrix H

LOW, IGH   Integers typically determined by the balancing subroutine
            BALANC. If BALANC is not used, set LOW = 1 and
            IGH = N.

H   Matrix in upper Hessenberg form stored in a real two-
            dimensional array. Information about the transformations
            used in the reduction to Hessenberg form by subroutine
            ELMHES, if performed, is stored in the remaining triangle
            under the Hessenberg matrix. Upon normal return, H is
            destroyed.

c. Output Arguments

WR, WI   One-dimensional arrays, dimensioned at least N in the
            calling program, containing, upon normal return, the real
            and imaginary parts of the eigenvalues, respectively. The
            eigenvalues are unordered except that the complex conjugate

160

pairs of values appear consecutively with the eigenvalue
having positive imaginary part first.

IERR   Integer error code:

    IERR = 0   Normal return

    IERR = J   The Jth eigenvalue has not been determined
               after 30 iterations of the QR algorithm.
               If an error exit is made, the eigenvalues
               should be correct for indices
               IERR+1,IERR+2,...,N.

d. COMMON Blocks

None

e. Error Messages

None directly from HQR; IERR should be examined upon return.

f. Subroutines Employed by HQR

None

g. Subroutine Employing HQR

EIGEN

h. Concluding Remarks

None

E/ Eigenvectors (INVIT)

1. PURPOSE

Subroutine INVIT finds those eigenvectors of a real square upper
Hessenberg matrix corresponding to specified eigenvalues using inverse
iteration. The subroutine INVIT is a translation of the ALGOL procedure
INVIT found on pages 418 to 439 of [6-4].

2. USAGE

a. Calling Sequence

CALL INVIT(NM,N,A,WR,WI,SELECT,MM,M,Z,IERR,RM1,RV1,RV2)

b. Input Arguments

NM	First dimension of the array A as given in the calling program
N	Number of rows in A
A	Matrix in upper Hessenberg form stored in real two-dimensional array. Upon return, A is unaltered.
WR, WI	One-dimensional arrays dimensioned at least N by the calling program. WR and WI contain the real and imaginary parts, respectively, of the eigenvalues of the matrix. The eigenvalues must be stored in a manner identical to that of subroutine HQR. WI is unaltered upon return. WR may be altered since close eigenvalues are perturbed slightly in searching for independent eigenvectors.
SELECT	One-dimensional array of logical variables, dimensioned at least N by the calling program. SELECT specifies the eigenvectors to be found. The eigenvector corresponding to the Jth eigenvalue is specified by setting SELECT(J) to TRUE. Upon return, SELECT may be altered. If the elements corresponding to a pair of conjugate complex eigenvalues were each initially set TRUE, the program resets the second of the two elements to FALSE.
MM	MM should be set to an upper bound for the number of columns required to store the eigenvectors to be found. Note that two columns are required to store the eigenvector corresponding to a complex eigenvalue.

162

RM1, RV1, RV2   Temporary storage arrays dimensioned at least N × N,

                  N, and N, respectively, by the calling program

c. Output Arguments

M   The actual number of columns used to store the

                  eigenvectors

Z   Matrix dimensioned at least NM × MM by the calling

                  program. Z contains the real and imaginary parts of

                  the eigenvectors. If the next selected eigenvalue is

                  real, the next column of Z contains its eigenvector.

                  If the eigenvalue is complex, the next two columns

                  of Z contain the real and imaginary parts of its

                  eigenvector. The eigenvectors are normalized so that

                  the component of largest magnitude is unity. Any

                  vector which fails the acceptance test is set to zero.

IERR   Integer error code:

                  IERR = 0   Normal return

                  IERR = -(2N + 1) More than MM columns of Z are

                                    necessary to store the eigen-

                                    vectors corresponding to the

                                    specified eigenvalues.

                  IERR = -K   The iteration corresponding to the

                                    Kth value fails.

                  IERR = -(N + K)   Both of the above error situations

                                    occur.

d. COMMON Blocks

None

e. Error Messages

None directly from INVIT; IERR should be examined upon return.

163

f. Subroutines Employed by INVIT

None

g. Subroutine Employing INVIT

EIGEN

h. Concluding Remarks

None

F/ Eigenvectors (ELMBAK)

1. PURPOSE

Subroutine ELMBAK forms the eigenvectors of a real square matrix A by back transforming those of the corresponding upper Hessenberg matrix determined by subroutine ELMHES. ELMBAK is a translation of the ALGOL procedure ELMBAK found on pages 339 to 358 of [6-4].

2. USAGE

a. Calling Sequence

CALL ELMBAK(NM,LOW,IGH,A,INT,M,Z)

b. Input Arguments

NM   First dimension of the array A as given in the calling
            program

LOW, IGH   Integers available from the balancing subroutine BALANC. If
            BALANC is not used, set LOW = 1 and IGH = The order of
            the matrix.

A   Two-dimensional array dimensioned at least NM × IGH in the
            calling program. A contains the multipliers which were
            used in the reduction by ELMHES in its lower triangle below
            the subdiagonal.

INT   One-dimensional array dimensioned at least IGH by the calling
            program. INT contains information on the rows and columns
            interchanged in the reduction by ELMHES. Only elements LOW
            through IGH are used.

M   The number of columns of Z to be back transformed

Z   Two-dimensional array dimensioned at least NM × M by the

           calling program. Z contains the real and imaginary parts

           of the eigenvectors to be back transformed in its first

           M columns. The contents of Z are destroyed upon return.

c. Output Arguments

Z   The real and imaginary parts of the transformed eigenvectors

           are returned in the first M columns.

d. COMMON Blocks

None

e. Error Messages

None directly from ELMBAK

f. Subroutines Employed by ELMBAK

None

g. Subroutine Employing ELMBAK

EIGEN

h. Concluding Remarks

None

G/ Eigenvectors (BALBAK)

1. PURPOSE

Subroutine BALBAK forms the eigenvectors of a real square matrix by back
transforming those of the corresponding balanced matrix determined by
subroutine BALANC. BALBAK is a translation of the ALGOL procedure
BALBAK found on pages 315 to 326 of [6-4].

2. USAGE

a. Calling Sequence

CALL BALBAK(NM,N,LOW,IGH,SCALE,M,Z)

b. Input Arguments

NM   First dimension of the matrix array as given in the calling
              program

N   Number of rows of the matrix

LOW, IGH   Integers determined by BALANC

SCALE   One-dimensional array dimensioned at least N by the calling
              program. SCALE contains information determining the permu-
              tations and scaling factors used by BALANC.

M   The number of columns of Z to be back transformed

Z   Two-dimensional array dimensioned at least NM × M in the
              calling program. Z contains the real and imaginary parts
              of the eigenvectors to be back transformed in its first
              M columns.

c. Output Arguments

Z   The real and imaginary parts of the transformed eigenvectors
              are returned in the first M columns.

d. COMMON Blocks

None

e. Error Messages

None directly from BALBAK

f. Subroutines Employed by BALBAK

None

g. Subroutine Employing BALBAK

EIGEN

h. Concluding Remarks

None

H/ LU Factorization (DETFAC)

1. PURPOSE

Subroutine DETFAC factors a real square matrix A as

$$PA = LU \qquad\qquad (6\text{-}1)$$

where P is a permutation matrix representing row pivotal strategy,
L is a unit lower triangular matrix, and U is an upper triangular
matrix. Options are provided to compute the determinant of A with and
without A input in factored form.

2. USAGE

a. Calling Sequence

CALL DETFAC(NMAX,N,A,IPIVOT,IDET,DETERM,ISCALE,WK,IERR)

b. Input Arguments

NMAX
: First dimension of the array A as given in the calling program

N
: Number of rows of matrix A

A
: Two-dimensional array dimensioned at least NMAX × N in the calling program. A is the matrix to be factored. If the factored form of A is input,

$$A = (L\backslash U) \qquad\qquad (6\text{-}2)$$

should be used neglecting the unity elements of L and the pivotal strategy input through the array IPIVOT. For A in unfactored form, input data are destroyed.

IPIVOT
: One-dimensional array dimensioned at least N by the calling program. Not required as input if the unfactored form of A is used. Otherwise, IPIVOT(I) = J indicates that row J of matrix A was used to pivot for the Ith column.

IDET   Determinant evaluation code:

                 0   Compute L and U matrices only.

                 1   Given L and U matrices as input, compute
                      parameters defining the determinant.

                 2   Compute L, U, and determinant parameters.

WK   One-dimensional array dimensioned at least N by the
                 calling program and used as a work storage array.

c. Output Arguments

A   Upon normal return, the L and U matrices are over
                 stored in A as

$$A = (L\backslash U) \qquad\qquad (6\text{-}3)$$

                 neglecting the unity elements in L.

IPIVOT   Upon normal return, IPIVOT contains the pivotal
                 strategy as previously explained.

DETERM, ISCALE   Determinant evaluation parameters; upon return, for
                 IDET $\neq$ 0,
                 $$\det(A) = \text{DETERM} \times 10^{100 \times \text{ISCALE}}$$

IERR   Singularity test parameter:

                 0   Matrix A is singular.

                 1   Matrix A is nonsingular.

d. COMMON Blocks

None

e. Error Messages

None directly from DETFAC; IERR should be examined upon return.

f. Subroutines Employed by DETFAC

None

g. Subroutine Employing DETFAC

GELIM

h. Concluding Remarks

None

I/ Solve AX + XB = C (AXPXB)

1. PURPOSE

Subroutine AXPXB solves the real matrix equation

$$AX + XB = C \tag{6-4}$$

where A, B, and C are constant matrices of order $m \times m$, $n \times n$, and $m \times n$, respectively. The matrices A and B are transformed into real lower and upper Schur form [6-2], and the transformed system is solved by back substitution. The option is provided to input the Schur forms directly and bypass the Schur decomposition.

2. USAGE

a. Calling Sequence

CALL AXPXB(A,U,M,NA,NU,B,V,N,NB,NV,C,NC,EPSA,EPSB,FAIL)

b. Input Arguments

A   Two-dimensional array dimensioned at least $(m + 1) \times (m + 1)$ in the calling program. The upper $m \times m$ part contains the matrix A. If the Schur form of A is input directly the lower triangle and superdiagonal of the upper $m \times m$ part of the array A contain a lower Schur form of A.

U   Two-dimensional array dimensioned at least $m \times m$ in the calling program. Not required as input if the Schur form of A is not input directly. Otherwise, U contains the orthogonal matrix that reduces A to Schur form $\tilde{A}$ via

$$\tilde{A} = U'AU \tag{6-5}$$

M   Number of rows of the matrix A

NA First dimension of the array A; at least $m + 1$

NU First dimension of the array U; at least m

B   Two-dimensional array dimensioned at least $(n + 1) \times (n + 1)$
            in the calling program. The upper $n \times n$ part contains the
            matrix B. If the Schur form of B is input directly, the
            upper triangle and subdiagonal of the upper $n \times n$ part of
            array B contain an upper real Schur form of B.

V   Two-dimensional array dimensioned at least $n \times n$ in the call-
            ing program. Not required as input if the Schur form of A
            is not input directly. Otherwise, V contains the orthogonal
            matrix that reduces B to Schur form $\tilde{B}$ via

$$\tilde{B} = V'BV \qquad\qquad (6\text{-}6)$$

N   Number of rows of the matrix B

NB   First dimension of the array B; at least $n + 1$

NV   First dimension of the array V; at least n

C   Two-dimensional array dimensioned at least $m \times n$ by the
            calling program; destroyed upon return

NC   First dimension of the array C; at least m

EPSA   Convergence criterion for the reduction of A to Schur form.
            EPSA should be set slightly smaller than 10^{-k_a} where k_a
            is the number of significant digits in the elements of A.
            Set EPSA negative if a Schur form for A and transformation
            matrix U are directly input.

EPSB   Convergence criterion for the reduction of B to Schur form.
            EPSB should be set slightly smaller than 10^{-k_b} where k_b
            is the number of significant digits in the elements of B.
            Set EPSB negative if a Schur form for B and transformation
            matrix V are directly input.

c. Output Arguments

A   Upon normal return, the array A contains the lower Schur

                form of the matrix A.

U   Upon normal return, the array U contains the orthogonal

                matrix U which transforms A to Schur form.

B   Upon normal return, the array B contains the upper Schur form

                for the matrix B.

V   Upon normal return, the array V contains the orthogonal

                matrix V which transforms B to upper Schur form.

C   Upon normal return, the solution X is stored in C.

FAIL   Integer variable containing an error signal. If FAIL is posi-

                tive (negative) then the program was unable to reduce A(B)

                to real Schur form. If FAIL = 0, the reductions proceeded

                without mishap.

d. COMMON Blocks

None

e. Error Messages

None directly from AXPXB; FAIL should be tested upon return.

f. Subroutines Employed by AXPXB

HSHLDR, BCKMLT, SCHUR, SHRSLV

g. Subroutine Employing AXPXB

BARSTW

h. Concluding Remarks

Subroutine AXPXB should be used when AX + XB = C needs to be solved

for a number of C matrices. For the first C, set EPSA and EPSB based

on k_a and k_b. Afterward, assuming FAIL = 0, set EPSA and EPSB to

negative values and compute X for the remaining C matrices. For the special case in which

$$A = B' \quad \text{and} \quad C = C' \tag{6-7}$$

the subroutine ATXPXA should be employed.

J/ Solve AX + XB = C (SHRSLV)

1. PURPOSE

Subroutine SHRSLV solves the real matrix equation

$$AX + XB = C \tag{6-8}$$

where A is an $m \times m$ matrix in lower real Schur form and B is an
$n \times n$ matrix in upper real Schur form.

2. USAGE

a. Calling Sequence

CALL SHRSLV(A,B,C,M,N,NA,NB,NC)

b. Input Arguments

A   Two-dimensional array dimensioned at least $m \times m$ in the calling program. A contains the lower Schur form of the
matrix A. Not destroyed upon return.

B   Two-dimensional array dimensioned at least $n \times n$ in the calling program. B contains the matrix B in upper Schur form.
Not destroyed upon return.

C   Two-dimensional array dimensioned at least $m \times n$ by the calling program. C contains the C matrix of the algebraic
equation which is destroyed upon return.

M   Number of rows of the matrix A

N   Number of rows of the matrix B

NA   First dimension of the array A; at least m

NB   First dimension of the array B; at least n

NC   First dimension of the array C; at least m

c. Output Arguments

C   Upon normal return, the solution X is in C.

d. COMMON Blocks

SLVBLK

e. Error Messages

None directly from SHRSLV

f. Subroutine Employed by SHRSLV

SYSSLV

g. Subroutine Employing SHRSLV

AXPXB

h. Concluding Remarks

COMMON block SLVBLK is internal to subroutines SHRSLV, SYMSLV, and
SYSSLV. It requires no input by the user. Contents of SLVBLK are
given in the description of subroutine SYSSLV.

K/ Solve A'X + XA = C (ATXPXA)

1. PURPOSE

Subroutine ATXPXA solves the real matrix equation

$$A'X + XA = C \tag{6-9}$$

where A and C are constant matrices of dimension $n \times n$ with

$$C = C' \tag{6-10}$$

The matrix A is transformed into upper Schur form [6-2] and the transformed system is solved by back substitution. The option is provided to
input the Schur form directly and bypass the Schur decomposition.

173

2. USAGE

a. Calling Sequence

CALL ATXPXA(A,U,C,N,NA,NU,NC,EPS,FAIL)

b. Input Arguments

A   Two-dimensional array dimensioned at least $(n + 1) \times (n + 1)$ in the calling program. The upper $n \times n$ part contains the matrix A. If the Schur form of A is input directly the upper triangle and the first subdiagonal of the upper $n \times n$ part of the array A contain an upper real Schur form of A.

U   Two-dimensional array dimensioned at least $n \times n$ in the calling program. Not required as input if the Schur form of A is not input directly. Otherwise, U contains the orthogonal matrix that reduces A to Schur form $\tilde{A}$ via

$$\tilde{A} = U'AU \qquad\qquad (6\text{-}11)$$

C   Two-dimensional array dimensioned at least $n \times n$ by the calling program; destroyed upon return

N   Number of rows of the matrix A

NA First dimension of the array A; at least $n + 1$

NU First dimension of the array U; at least n

NC First dimension of the array C; at least n

EPS Convergence criterion for the reduction of A to Schur form. EPS should be set slightly smaller than 10^{-k_a} where k_a is the number of significant digits in the elements of A. Set EPS negative if a Schur form for A and transformation matrix U are directly input.

c. Output Arguments

A   Upon normal return, the array A contains the upper Schur form of the matrix A. Same as input if EPS < 0.

U   Upon normal return, the array U contains the orthogonal

                matrix U which transforms A to Schur form. Same as input

                if EPS < 0.

C   Upon normal return, the solution X is stored in C.

FAIL   Integer variable containing an error signal. If FAIL is non-

                zero, the program was unable to reduce A to real Schur form.

                If FAIL = 0, the reduction proceeded without mishap.

d. COMMON Blocks

None

e. Error Messages

None directly from ATXPXA. FAIL should be tested upon return.

f. Subroutines Employed by ATXPXA

HSHLDR, BCKMLT, SCHUR, SYMSLV

g. Subroutine Employing ATXPXA

BARSTW

h. Concluding Remarks

Subroutine ATXPXA should be used when

$$A'X + XA = C \qquad\qquad (6\text{-}9)$$

needs to be solved for a number of symmetric C matrices. For the

first C, set EPS based on k_a. Afterward, assuming FAIL = 0, set EPS

negative and compute X for the remaining C matrices.

L/ Solve A'X + XA = C (SYMSLV)

1. PURPOSE

Subroutine SYMSLV solves the real matrix equation

$$A'X + XA = C \qquad\qquad (6\text{-}12)$$

where C = C' and A is n × n and in upper real Schur form.

2. USAGE

a. Calling Sequence

CALL SYMSLV(A,C,N,NA,NC)

b. Input Arguments

A   Two-dimensional array dimensioned at least n × n in the

     calling program. A contains the upper Schur form for the

     matrix A. Not destroyed upon return.

C   Two-dimensional array dimensioned at least n × n in the

     calling program. C contains the C matrix of the algebraic

     equation which is destroyed upon return.

N   Number of rows of the matrix A

NA First dimension of the array A; at least n

NC First dimension of the array C; at least n

c. Output Arguments

C   Upon normal return, the solution X is stored in C.

d. COMMON Blocks

SLVBLK

e. Error Messages

None directly from SYMSLV

f. Subroutine Employed by SYMSLV

SYSSLV

g. Subroutine Employing SYMSLV

ATXPXA

h. Concluding Remarks

COMMON block SLVBLK is internal to subroutines SHRSLV, SYMSLV, and

SYSSLV. It requires no input by the user. Contents of SLVBLK are

given in the description of subroutine SYSSLV.

M/ Upper Hessenberg Form (HSHLDR)

1. PURPOSE

Subroutine HSHLDR reduces a real $n \times n$ matrix A to upper Hessenberg form by Householder's method of elementary Hermitian transformations described in [6-7].

2. USAGE

a. Calling Sequence

CALL HSHLDR(A,N,NA)

b. Input Arguments

A Two-dimensional array dimensioned at least $(n + 1) \times (n + 1)$ in the calling program. The upper $n \times n$ part of the array A contains the matrix A which is destroyed upon return.

N Number of rows of matrix A

NA First dimension of the array A; at least $n + 1$

c. Output Arguments

A Upon normal return, the upper triangle of the array A to column n contains the upper triangle of the Hessenberg form of the matrix A. Column $n + 1$ contains the subdiagonal elements of the Hessenberg form. The lower triangle and the $(n + 1)$th row of the array contain a history of the Householder transformations.

d. COMMON Blocks

None

e. Error Messages

None directly from HSHLDR

f. Subroutines Employed by HSHLDR

None

g. Subroutines Employing HSHLDR

AXPXB, ATXPXA

h. Concluding Remarks

None

N/ Upper Hessenberg Form (BCKMLT)

1. PURPOSE

Given the output A from subroutine HSHLDR, subroutine BCKMLT computes
the orthogonal matrix that reduces A to upper Hessenberg form.

2. USAGE

a. Calling Sequence

CALL BCKMLT(A,U,N,NA,NU)

b. Input Arguments

A   Two-dimensional array containing the output from HSHLDR

N   Number of rows in matrix A in HSHLDR

NA   First dimension of the array A; at least N + 1

NU   First dimension of the array U; at least N

c. Output Arguments

U   Two-dimensional array dimensioned at least n × n where n is
    the order of the A matrix in HSHLDR. If the matrix A is
    used for U in the calling sequence, the elements of the
    orthogonal matrix will overwrite the output of HSHLDR.

d. COMMON Blocks

None

e. Error Messages

None directly from BCKMLT

f. Subroutines Employed by BCKMLT

None

g. Subroutines Employing BCKMLT

AXPXB, ATXPXA

h. Concluding Remarks

None

O/ Real Schur Form (SCHUR)

1. PURPOSE

Subroutine SCHUR reduces an $n \times n$ upper Hessenberg matrix H to real
Schur form. Computation is by the QR algorithm with implicit origin
shifts. The program SCHUR is an adaptation of the ALGOL program HQR
found in [6-6]. The product of the transformations used in the reduction
is accumulated.

2. USAGE

a. Calling Sequence

CALL SCHUR(H,U,NN,NH,NU,EPS,FAIL)

b. Input Arguments

H   Two-dimensional array dimensioned at least $n \times n$ in the
             calling program. The array H contains the matrix H in
             upper Hessenberg form. The elements below the third sub-
             diagonal are undisturbed upon return.

U   Two-dimensional array dimensioned at least $n \times n$ by the
             calling program. On input, U contains any square matrix
             desired.

NN   The number of rows in the matrices H and U

NH   First dimension of the array H; at least n

NU   First dimension of the array U; at least n

EPS   Number used in determining when an element of H is negligible.
             The element $H(i,j)$ is negligible if $|H(i,j)| \leq EPS \times \|H\|$
             where $\|H\|$ denotes the ℓ_∞ norm of H.

179

c. Output Arguments

H   Upon normal return, H contains an upper Schur form of H.

U   Upon normal return, U contains the product of the input U

    right multiplied by the accumulated orthogonal transforma-

    tions used to reduce H to Schur form. If the identity

    matrix is input for U, then the Schur form is

       $U'HU$

FAIL Integer variable containing an error signal. If FAIL is

    positive, then the program failed to make the (FAIL-1) or

    (FAIL-2) subdiagonal element negligible after 30 iterations

    of the QR algorithm.

d. COMMON Blocks

None

e. Error Messages

None directly from SCHUR

f. Subroutines Employed by SCHUR

None

g. Subroutines Employing SCHUR

AXPXB, ATXPXA

h. Concluding Remarks

None

P/ Solve Ax = b (SYSSLV)

1. PURPOSE

Subroutine SYSSLV solves the linear system

   $Ax = b$ (6-13)

where A is an $n \times n$ $(n \leq 5)$ matrix and b is an n-dimensional

vector. Solution is by Crout reduction. The matrix A, the vector b,

and order n are contained in the arrays A, B, and the variable N
of the COMMON block SLVBLK. The solution is returned in the array B.

2. USAGE

a. Calling Sequence

CALL SYSSLV

b. Input Arguments

None

c. Output Arguments

None

d. COMMON Blocks

SLVBLK

e. Error Messages

None directly from SYSSLV

f. Subroutines Employed by SYSSLV

None

g. Subroutines Employing SYSSLV

SHRSLV, SYMSLV

h. Concluding Remarks

None

Q/ Solve AX = B (GAUSEL)

1. PURPOSE

Subroutine GAUSEL solves a set of linear equations,

$$AX = B \qquad\qquad (6\text{-}14)$$

by the method of Gaussian elimination. The constant matrices A and B
are of dimension $n \times n$ and $n \times r$, respectively. No information is
returned on the pivotal strategy or the value of the determinant of A.

2. USAGE

a. Calling Sequence

CALL GAUSEL(MAX,N,A,NR,B,IERR)

b. Input Arguments

MAX   First dimension of the array B given in the calling program

N   Number of rows of the matrix A

A   Two-dimensional array dimensioned at least $n \times n$ in the calling program; destroyed upon return

NR   Number of columns in the matrix B

B   Two-dimensional array dimensioned at least $MAX \times NR$ in the calling program; destroyed upon return

c. Output Arguments

B   Upon normal return, B contains the solution X.

IERR   Integer error code:

     0   Normal return

     2   Input matrix A is singular.

d. COMMON Blocks

None

e. Error Messages

None directly from GAUSEL. IERR should be checked upon return.

f. Subroutines Employed by GAUSEL

None

g. Subroutine Employing GAUSEL

EXPADE

h. Concluding Remarks

None

REFERENCES

6-1. J. S. White and H. Q. Lee, <u>Users Manual for the Variable Dimension</u> <u>Automatic Synthesis Program (VASP)</u>, NASA TM X-2417, 1971.

6-2. R. H. Bartels and G. W. Stewart, "Algorithm 432 - Solution of the Matrix Equation AX + XB = C," <u>Commun. ACM</u>, <u>vol. 15</u>, no. 9, Sept. 1972, pp. 820-826.

6-3. B. N. Parlett and C. Reinsch, "Balancing a Matrix for Calculation of Eigenvalues and Eigenvectors," <u>Numer. Math.</u>, <u>Bd. 13</u>, Heft 4, 1969, pp. 293-304.

6-4. J. H. Wilkinson and C. Reinsch, <u>Handbook for Automatic Computation</u>, <u>Volume II - Linear Algebra</u>, Springer-Verlag, 1971.

6-5. R. S. Martin and J. H. Wilkinson, "Similarity Reduction of a General Matrix to Hessenberg Form," <u>Numer. Math.</u>, <u>Bd. 12</u>, Heft 5, 1968, pp. 349-368.

6-6. R. S. Martin, G. Peters, and J. H. Wilkinson, "The QR Algorithm for Real Hessenberg Matrices," <u>Numer. Math.</u>, <u>Bd. 14</u>, Heft 3, 1970, pp. 219-231.

6-7. J. H. Wilkinson, <u>The Algebraic Eigenvalue Problem</u>, Clarendon Press (Oxford), 1965.

SEVEN/ DESIGN PROBLEMS

I/ INTRODUCTION

This chapter contains selected problems illustrating the use of the ORACLS program subroutines. Sample executive programs and output data are presented which develop state variable feedback control laws using the optimal transient regulator, optimal sampled-data regulator, and model-following design approaches. Additionally, the construction of an asymptotic Kalman-Bucy estimator, the eigenvalue placement procedure, and the calculation of a transfer matrix for a constant linear system are illustrated.

Data employed for the plant equations are from linearized mathematical models of the lateral and longitudinal dynamics of an F-8 aircraft [7-1]. The problem and machine-dependent accuracy and convergence parameters required for COMMON blocks TOL and CONV of subroutine RDTITL used in the example computations are:

$$
\begin{aligned}
&\text{EPSAM = EPSBM = 1.E-10} \\
&\text{IACM = 12} \\
&\text{SUMCV = 1.E-8} \\
&\text{MAXSUM = 50} \\
&\text{RICTVC = 1.E-8} \\
&\text{SERCV = 1.E-8}
\end{aligned}
\qquad (7\text{-}1)
$$

The construction of RDTITL for the computations is shown in figure 7-1. All computations were performed using the Control Data Cyber digital computer system under Network Operating System (NOS) 1.3.

```
      SUBROUTINE RDTITL
C
C
C     A CALL TO RDTITL SHOULD PRECEDE CALLS TO OTHER ORACLS
C     SUBROUTINES BY THE EXECUTIVE PROGRAM
C
C
C
      COMMON/LINES/NLP,LIN,TITLE(8),TIL(2)
      COMMON/FORM/NEPR,FMT1(6),FMT2(6)
      COMMON/TOL/EPSAM,EPSBM,IACM
      COMMON/CONV/SUMCV,MAXSUM,RICTCV,SERCV
C     NLP = NO. LINES/PAGE VARIES WITH THE INSTALLATION
      DATA LIN,NLP/1,44/
      DATA NEPR,FMT1/7,10H(1P7E16.7)/
      DATA TIL/10H      ORACL,10HS  PROGRAM/
      DATA FMT2/10H(3X,1P7E16,10H.7)
      DATA EPSAM/1.E-10/
      DATA EPSBM/1.E-10/
      DATA IACM/12/
      DATA SUMCV/1.E-8/
      DATA RICTCV/1.E-8/
      DATA SERCV/1.E-8/
      DATA MAXSUM/50/
      READ(5,100) TITLE
      IF(EOF(5))90,91
   90 CONTINUE
      STOP 1
   91 CONTINUE
  100 FORMAT(8A10)
      CALL LNCNT(100)
      RETURN
      END
```

FIG. 7-1 Subroutine RDTITL for the design problems.

II/ OPTIMAL TRANSIENT REGULATOR'

A/ Problem Statement

This problem illustrates the construction of storage arrays for ORACLS
and demonstrates the solution of a simple transient optimal linear
regulator problem. Given the system

$$\dot{x}(t) = A\ x(t) + B\ u(t) \tag{7-2}$$

with

$$A = \begin{bmatrix} -2.60 & 0.25 & -38.0 & 0.0 \\ -0.075 & -0.27 & 4.40 & 0.0 \\ 0.078 & -0.99 & -0.23 & 0.052 \\ 1.0 & 0.078 & 0.00 & 0.00 \end{bmatrix} \tag{7-3}$$

and

$$B = \begin{bmatrix} 17. & 7.0 \\ 0.82 & -3.2 \\ 0.0 & 0.046 \\ 0.0 & 0.0 \end{bmatrix} \tag{7-4}$$

find the control law $u(t) = -F(t)\ x(t)$ which minimizes

$$J = \frac{1}{2}\ x'(20)\ x(20) + \int_0^{20} \left[x'(t)\ x(t) + 100\ u'(t)\ u(t) \right]\ dt \tag{7-5}$$

The solution can be found by directly employing the subroutine CNTNREG.
Output data are to be printed at 2.0-second intervals between 0 and 20.

187

Referring to the description of CNTNREG, the dimensions of the subroutine argument arrays must be specified. The matrices A, B, Q, and R are 4×4, 4×2, 4×4, and 2×2, respectively. As packed one-dimensional arrays, they must be diminished at least A(16), B(8), Q(16), and R(4). However, for this example, input data to CNTNREG are defined within the source program and, for convenience of construction, two-dimensional arrays A(4,4), B(4,2), Q(4,4), and R(2,2) are used. Since $H = I_4$, no data for H are required if the logical variable IDENT is set to TRUE. The array P initially contains the weighting matrix $\frac{1}{2} I_4$ on the final states and hence, is dimensioned P(16). The array F will contain the feedback gain matrix and must be dimensioned at least F(8). Similarly, Z, W, LAMBDA, and S are dimensioned 64, 64, 16, and 16, respectively. Dimensions for NA, NB, NQ, NR, NF, and NP are set to 2. The vector T contains the final problem time and print interval and is also dimensioned 2. The vector IOP controls the printing from CNTNREG and must be dimensioned at least 3. The dimension of DUMMY is computed from the formula (since IOP(3) = 0 for a transient solution to the Riccati equation),

$$9n^2 + 17n + nr = \text{dim. DUMMY} \qquad (7\text{-}6)$$

with $n = 4$ and $r = 2$.

The following executive program shows how the specific data for CNTNREG may be constructed. Note that subroutines NULL and UNITY may be employed to generate zero and unity elements in the coefficient and weighting matrices. For the NOS 1.3 system at LaRC tape 5 is designated input and tape 6 output, as indicated by the source program card. Output from ORACLS follows the executive program.

B/ Executive Program

```
1          PROGRAM REGLAT(INPUT,OUTPUT,TAPE5=INPUT,TAPE6=OUTPUT)
     C******************************************************************
     C*
     C*
     C*
5    C*     EXECUTIVE PROGRAM FOR TRANSIENT REGULATOR PROBLEM
     C*
     C*
     C******************************************************************
           DIMENSION A(4,4),B(4,2),Q(4,4),R(2,2),Z(64),W(64),LAMBDA(16),S(16)
10        1,F(2,4),P(16),DUMMY(220),NA(2),NB(2),NQ(2),NR(2),IOP(3),T(2),NF(2)
          2,NP(2)
           LOGICAL IDENT
           REAL LAMBDA
15   C
     C
     C          INITIALIZE   NA,NB,...,NP
           NA(1)= 4
20         NA(2)= 4
           NB(1)= 4
           NB(2)= 2
           NQ(1)= 4
           NQ(2)= 4
25         NR(1)= 2
           NR(2)= 2
           NP(1)= 4
           NP(2)= 4
     C
30   C          SET FINAL TIME AND PRINT INTERVAL
           T(1)= 20.
           T(2)= 2.
     C
     C          DEFINE PRINT AND TRANSIENT SOLUTION OPTIONS
35         IOP(1)= 1
           IOP(2)= 1
           IOP(3)= 0
     C
     C          DEFINE COEFFICIENT AND WEIGHTING MATRICES
40         CALL NULL(A,NA)
           CALL NULL(B,NB)
           CALL UNITY(P,NP)
```

189

```
            CALL SCALE(P,NP,P,NP,0.5)
            CALL UNITY(Q,NQ)
            CALL UNITY(R,NR)
            CALL SCALE(R,NR,R,NR,100.)
      C
45          A(1,1)= -2.6
            A(1,2)=  0.25
            A(1,3)= -38.
            A(2,1)= -.075
            A(2,2)= -.27
            A(2,3)=  4.4
50          A(3,1)=  0.078
            A(3,2)= -.99
            A(3,3)= -.23
            A(3,4)=  0.052
            A(4,1)=  1.0
55          A(4,2)=  0.078
            B(1,1)=  17.0
            B(1,2)=  7.0
            B(2,1)=  0.82
            B(2,2)= -3.2
60          B(3,2)=  0.046
            IDENT= .TRUE.
      C
      C     INPUT HOLLERITH DATA FOR TITLE OF OUTPUT
65          CALL RDTITL
      C
      C     SOLVE TRANSIENT REGULATOR PROBLEM
70          CALL CNTNREG(A,NA,B,NB,H,NH,Q,NQ,R,NR,Z,W,LAMBDA,S,F,NF,P,NP,T,IOP
           1,IDENT,DUMMY)
      C
      C
75          STOP
            END
```

OUTPUT FOR TRANSIENT REGULATOR PROBLEM

PROGRAM TO SOLVE THE TIME-INVARIANT FINITE-DURATION CONTINUOUS OPTIMAL
REGULATOR PROBLEM WITH NOISE-FREE MEASUREMENTS

```
     A   MATRIX        4 ROWS        4 COLUMNS
-2.6000000E+00   2.5000000E-01  -3.8000000E+01   0.
-7.5000000E-02  -2.7000000E-01   4.4000000E+00   0.
 7.8000000E-02  -9.9000000E-01  -2.3000000E-01   5.2000000E-02
 1.0000000E+00   7.8000000E-02   0.              0.
```

```
     B   MATRIX        4 ROWS        2 COLUMNS
 1.7000000E+01   7.0000000E+00
 8.2000000E-01  -3.2000000E+00
 0.              4.6000000E-02
 0.              0.
```

```
     Q   MATRIX        4 ROWS        4 COLUMNS
 1.0000000E+00   0.              0.              0.
 0.              1.0000000E+00   0.              0.
 0.              0.              1.0000000E+00   0.
 0.              0.              0.              1.0000000E+00
```

H IS AN IDENTITY MATRIX

```
     R   MATRIX        2 ROWS        2 COLUMNS
 1.0000000E+02   0.
 0.              1.0000000E+02
```

WEIGHTING ON TERMINAL VALUE OF STATE VECTOR

```
     P   MATRIX        4 ROWS        4 COLUMNS
 5.0000000E-01   0.              0.              0.
 0.              5.0000000E-01   0.              0.
 0.              0.              5.0000000E-01   0.
 0.              0.              0.              5.0000000E-01
```

```
     Z   MATRIX        8 ROWS        8 COLUMNS
```

191

ORACLS PROGRAM

```
-2.6000000E+00   2.5000000E-01  -3.8000000E+01   0.             -3.3800000E+00   8.4600000E-02  -3.2200000E-03
0.
-7.5000000E-02  -2.7000000E-01   4.4000000E+00   0.              8.4600000E-02  -1.0912400E-01   1.4720000E-03
0.
 7.8000000E-02  -9.9000000E-01  -2.3000000E-01   5.2000000E-02  -3.2200000E-03   1.4720000E-03  -2.1160000E-05
0.
 1.0000000E+00   7.8000000E-02   0.              0.              0.              0.              0.
0.
-1.0000000E+00   0.              0.              0.              2.6000000E+00   7.5000000E-02  -7.8000000E-02
-1.0000000E+00
 0.             -1.0000000E+00   0.              0.             -2.5000000E-01   2.7000000E-01   9.9000000E-01
-7.8000000E-02
 0.              0.             -1.0000000E+00   0.              3.8000000E+01  -4.4000000E+00   2.3000000E-01
0.
 0.              0.              0.             -1.0000000E+00   0.              0.             -5.2000000E-02
0.
```

EIGENVALUES OF Z

```
-6.4605654E-01   0.
 6.4605654E-01   0.
 7.8466995E-01   2.6501143E+00
 7.8466995E-01  -2.6501143E+00
-7.8466995E-01   2.6501143E+00
-7.8466995E-01  -2.6501143E+00
-2.8451457E+00   0.
 2.8451457E+00   0.
```

CORRESPONDING EIGENVECTORS

```
-2.4733021E-01  -1.7233188E-01   5.0257395E-02   2.6582832E-02  -8.5528327E-02  -1.6423055E-01   6.2134586E-01
-6.2618218E-02
 4.3326566E-03  -3.3791761E-02  -1.6344167E-02   9.5183892E-03   3.5983643E-02   1.2843065E-02   3.0404804E-02
-6.9930617E-03
 7.5286043E-03   7.7409397E-03   1.7516130E-06  -8.0834088E-03  -4.9824656E-03   1.7745854E-02  -2.6988189E-03
 3.3935252E-04
 3.8230750E-01  -2.7082403E-01   1.4511446E-02  -1.4186503E-02  -4.8131275E-02   4.5465498E-02  -2.1922162E-01
-2.2200507E-02
```

192

OUTPUT FOR TRANSIENT REGULATOR PROBLEM

                                                                    DRACLS PROGRAM

7.2555584E-02   9.5507170E-02 -2.2377640E-02   2.4483836E-02 -1.5780566E-02 -4.4315091E-02   8.4269310E-02
1.0441627E-01
5.3890433E-01   7.9262267E-01   3.3965417E-01 -7.8633601E-03   3.2328081E-01 -7.4606776E-03   2.3749420E-01
3.4982534E-01
-4.3207089E-01 3.2205187E-01   1.7435213E-01   9.2129547E-01 -2.9506394E-01   8.7411594E-01 -7.0239213E-01
9.2853245E-01
5.5697883E-01   3.9327414E-01 -1.4120636E-01   4.7159165E-03 -3.8062689E-02 -1.2681699E-02 -8.9885488E-02
-9.1676080E-03

REORDERED EIGENVECTORS

-1.7233188E-01   5.0257395E-02   2.6582832E-02 -6.2618218E-02 -2.4733021E-01 -1.6423055E-01 -8.5528327E-02
6.2134586E-01
-3.3791761E-02 -1.6344167E-02   9.5183892E-03 -6.9930617E-03   4.3326566E-03   1.2843065E-02   3.5983643E-02
3.0404804E-02
7.7409397E-03   1.7516130E-06 -8.0834088E-03   3.3935252E-04   7.5286043E-03   1.7745854E-02 -4.9824656E-03
-2.6988189E-03
-2.7084035E-01   1.4511446E-02 -1.4186503E-02 -2.2200507E-02   3.8230750E-01   4.5465498E-02 -4.8131275E-02
-2.1922162E-01
9.5507170E-02 -2.2377640E-02   2.4483836E-02   1.0441627E-01   7.2558584E-02 -4.4315091E-02 -1.5780566E-02
8.4269310E-02
7.9262267E-01   3.3965417E-01 -7.8633601E-03   3.4982534E-01   5.3890433E-01 -7.4606776E-03   3.2328081E-01
2.3749420E-01
3.2205187E-01   1.7435213E-01   9.2129547E-01 -7.0239213E-01   3.2205187E-01 -4.3207089E-01 -2.9506394E-01
8.7411594E-01
3.9327414E-01 -1.4120636E-02   4.7159165E-03 -9.1676080E-03   5.5697883E-01   5.5697883E-01   5.5697883E-01
-8.9885488E-02

LAMBDA MATRIX OF EIGENVALUES OF Z WITH POSITIVE REAL PARTS

6.4605654E-01   0.   0.   0.
0.   7.8466995E-01   2.6501143E+00 0.
0.   -2.6501143E+00   7.8466995E-01 0.
0.   0.   0.   2.8451457E+00

WIZW MATRIX   8 ROWS   8 COLUMNS

OUTPUT FOR TRANSIENT REGULATOR PROBLEM   ORACLS PROGRAM

6.4605654E-01   4.3539094E-13 -1.6913328E-13 -2.0390208E-13 -7.3247951E-14 -3.4396140E-13   8.6190507E-14
1.7480680E-13
9.9822886E-14   7.8466995E-01   2.6501143E+00 -4.8159991E-13 -8.2746769E-13   4.7260138E-13   1.5520196E-12
1.1224239E-13
1.5597072E-13 -2.6501143E+00   7.8466995E-01   2.6422617E-12 -1.7036632E-12   1.1402066E-12 -1.4882895E-12
-1.9369661E-12
-6.5446563E-14 -9.1129449E-13 -7.4776612E-13   2.8451457E+00 -1.3339999E-13   4.7238794E-14   6.1187571E-13
-1.2551633E-12
6.0377566E-14 -1.2656352E-12 -3.2722815E-13   1.4561670E-12 -6.4605654E-01   8.1668150E-13 -4.3938717E-13
-1.4901340E-12
1.3255405E-13 -3.0147102E-12 -1.4740753E-13   2.4210317E-12 -3.4414850E-13 -7.8466995E-01 -2.6501143E+00
-3.4247465E-12
-2.2830483E-13   1.3227021E-12   3.0763466E-12 -1.7386355E-12   1.0132695E-12   2.6501143E+00 -7.8466995E-01
9.5261412E-13
1.6908885E-13 -1.7696865E-12 -5.0533563E-13   2.5116453E-12 -2.5555875E-13   7.2909408E-13 -1.8733266E-12
-2.8451457E+00

W11 MATRIX   4 ROWS   4 COLUMNS
-1.7233188E-01   5.0257395E-02   2.6582832E-02 -6.2618216E-02
-3.3791761E-02 -1.6344167E-02   9.5183892E-03 -6.9930617E-03
7.7409397E-03   1.7516130E-06 -8.0834088E-03   3.3935252E-04
-2.7082403E-01   1.4511446E-02 -1.4186503E-02 -2.2205076E-02

W21 MATRIX   4 ROWS   4 COLUMNS
9.5071706E-02   -2.2377640E-02   2.4483836E-02   1.0441627E-01
7.9262267E-01   3.3965417E-01 -7.8633601E-03   3.4985346E-01
3.2205187E-01   1.7435213E-01   9.2129547E-01   9.2853245E-01
3.9327414E-01   -1.4120636E-02   4.7159165E-03 -9.1676080E-03

W12 MATRIX   4 ROWS   4 COLUMNS
-2.4733021E-01 -1.6423055E-02 -8.5528327E-02   6.2134586E-01
4.3326566E-03   1.2843065E-02   3.5983643E-02   3.0404804E-02
7.5286043E-03   1.7745854E-02 -4.9824656E-02 -2.6988189E-03
3.8230750E-01   4.5465498E-02 -4.8131275E-02 -2.1922162E-01

W22 MATRIX   4 ROWS   4 COLUMNS
7.2558584E-01   -4.4315091E-02 -1.5780566E-02   8.4269310E-02
5.3890433E-01   -7.4606776E-03   3.2328081E-01   2.3749420E-01
-4.3207089E-01   8.7411594E-01 -2.9506394E-01 -7.0239213E-01
5.5678783E-01   -1.2681699E-02 -3.8062689E-01 -8.9885488E-02

S MATRIX   4 ROWS   4 COLUMNS

OUTPUT FOR TRANSIENT REGULATOR PROBLEM

```
-6.9197483E-01   8.8040174E-02  -7.0617792E-03  -1.0579571E-01
 5.3014743E-01   1.1451662E-01  -4.2235758E-01  -1.6582063E+00
 1.0292420E+00  -6.8625132E-01   8.0290849E-01  -1.3835163E-01
-4.1882228E-01  -2.9967598E-01  -4.0340142E-01   1.2407188E+00
```

MATRIX (R INVERSE)X(B TRANSPOSE)

```
1.7000000E-01   8.2000000E-03   0.
7.0000000E-02  -3.2000000E-02   4.6000000E-04   0.
```

EXP(-LAMBDA X   .20000000E+01)

```
2.7468973E-01   0.              0.              0.
0.              1.1545067E-01   1.7323718E-01   0.
0.             -1.7323718E-01   1.1545067E-01   0.
0.              0.              0.              3.3786086E-03
```

TIME =   .20000000E+02

```
      P  MATRIX        4 ROWS          4 COLUMNS
5.0000000E-01   0.              0.              0.
0.              5.0000000E-01   0.              0.
0.              0.              5.0000000E-01   0.
0.              0.              0.              5.0000000E-01
```

TIME =   .18000000E+02

```
      P  MATRIX        4 ROWS          4 COLUMNS
 2.2480611E-01   3.6054050E-01  -1.4217206E+00   3.2023881E-01
 3.6054050E-01   1.0744607E+01  -8.8655022E+00   1.5409135E+00
-1.4217206E+00  -8.8655022E+00   4.5683890E+01  -2.6235118E+00
 3.2023881E-01   1.5409135E+00  -2.6235118E+00   1.5641109E+00
```

195

OUTPUT FOR TRANSIENT REGULATOR PROBLEM

```
     F    MATRIX        2 ROWS         4 COLUMNS
4.1173471E-02   1.4939767E-01  -3.1438962E-01   6.7076089E-02
3.5451406E-03  -3.2266773E-01   2.0519022E-01  -2.8099331E-02
```

TIME =   .16000000E+02

```
     P    MATRIX        4 ROWS         4 COLUMNS
2.3819322E-01   4.0985673E-01  -1.5275098E+00   3.6636522E-01
4.0985673E-01   1.0991594E+01  -9.0079421E+00   1.7136851E+00
-1.5275098E+00  -9.0079421E+00   4.9256328E+01  -2.9811826E+00
3.6636522E-01   1.7136851E+00  -2.9811826E+00   1.7228142E+00
```

```
     F    MATRIX        2 ROWS         4 COLUMNS
4.3853673E-02   1.5980672E-01  -3.3354180E-01   7.6334306E-02
2.8554559E-03  -3.2718471E-01   2.0398637E-01  -3.0563703E-02
```

TIME =   .14000000E+02

```
     P    MATRIX        4 ROWS         4 COLUMNS
2.3924330E-01   4.1302539E-01  -1.5297598E+00   3.6980853E-01
4.1302539E-01   1.1028965E+01  -9.0076508E+00   1.7284448E+00
-1.5297598E+00  -9.0076508E+00   4.9271952E+01  -2.9877059E+00
3.6980853E-01   1.7278448E+00  -2.9877059E+00   1.7346173E+00
```

```
     F    MATRIX        2 ROWS         4 COLUMNS
4.4058169E-02   1.6065183E-01  -3.3392190E-01   7.7035778E-02
2.8265289E-03  -3.2815862E-01   2.0382674E-01  -3.0778783E-02
```

TIME =   .12000000E+02

```
     P    MATRIX        4 ROWS         4 COLUMNS
2.3930520E-01   4.1337855E-01  -1.5297546E+00   3.7003289E-01
```

196

OUTPUT FOR TRANSIENT REGULATOR PROBLEM

```
 4.1337855E-01   1.1032206E+01  -9.0096823E+00   1.7293204E+00
-1.5297546E+01  -9.0096823E+00   4.9275911E+01  -2.9880218E+00
 3.7003289E-01   1.7293204E+00  -2.9880218E+00   1.7354617E+00

          F  MATRIX         2 ROWS         4 COLUMNS
 4.4071588E-02   1.6073844E-01  -3.3393768E-01   7.7086018E-02
 2.8195631E-03  -3.2823855E-01   2.0389393E-01  -3.0810439E-02
```

TIME =   .10000000E+02

```
          P  MATRIX         4 ROWS         4 COLUMNS
 2.3930917E-01   4.1341056E-01  -1.5297727E+00   3.7004881E-01
 4.1341056E-01   1.1032469E+01  -9.0098261E+00   1.7294493E+00
-1.5297727E+00  -9.0098261E+00   4.9276275E+01  -2.9880973E+00
 3.7004881E-01   1.7294493E+00  -2.9880973E+00   1.7355256E+00

          F  MATRIX         2 ROWS         4 COLUMNS
 4.4072525E-02   1.6074604E-01  -3.3394193E-01   7.7089781E-02
 2.8180087E-03  -3.2824478E-01   2.0389743E-01  -3.0813485E-02
```

TIME =   .80000000E+01

```
          P  MATRIX         4 ROWS         4 COLUMNS
 2.3930948E-01   4.1341291E-01  -1.5297749E+00   3.7005005E-01
 4.1341291E-01   1.1032489E+01  -9.0098419E+00   1.7294590E+00
-1.5297749E+00  -9.0098419E+00   4.9276293E+01  -2.9881061E+00
 3.7005005E-01   1.7294590E+00  -2.9881061E+00   1.7355306E+00

          F  MATRIX         2 ROWS         4 COLUMNS
 4.4072598E-02   1.6074661E-01  -3.3394244E-01   7.7090072E-02
 2.8187540E-03  -3.2824528E-01   2.0389779E-01  -3.0813714E-02
```

TIME =   .60000000E+01

197

OUTPUT FOR TRANSIENT REGULATOR PROBLEM

```
   P   MATRIX   4 ROWS        4 COLUMNS
2.3930951E-01  4.1341309E-01  -1.5297751E+00   3.7005015E-01
4.1341309E-01  1.1032491E+01  -9.0098433E+00   1.7294597E+00
-1.5297751E+00 -9.0098433E+00   4.9276294E+01  -2.9981068E+00
3.7005015E-01  1.7294597E+00  -2.9981068E+00   1.7355310E+00

   F   MATRIX   2 ROWS        4 COLUMNS
4.4072604E-02  1.6074665E-01  -3.394248E-01   7.7090094E-02
2.8187502E-03 -3.2824531E-01   2.0389783E-01  -3.0813729E-02
```

TIME = .40000000E+01

```
   P   MATRIX   4 ROWS        4 COLUMNS
2.3930951E-01  4.1341310E-01  -1.5297751E+00   3.7005015E-01
4.1341310E-01  1.1032491E+01  -9.0098434E+00   1.7294597E+00
-1.5297751E+00 -9.0098434E+00   4.9276294E+01  -2.9981068E+00
3.7005015E-01  1.7294597E+00  -2.9981068E+00   1.7355310E+00

   F   MATRIX   2 ROWS        4 COLUMNS
4.4072604E-02  1.6074665E-01  -3.394248E-01   7.7090096E-02
2.8187499E-03 -3.2824531E-01   2.0389783E-01  -3.0813730E-02
```

STEADY-STATE SOLUTION HAS BEEN REACHED IN CNTNREG

III/ OPTIMAL SAMPLED-DATA REGULATOR

A/ Problem Statement

This problem demonstrates the usual situation in which several ORACLS sub-routines are required to obtain a solution and input data are entered through use of the READ subroutine.

Given the linear time-invariant system

$$\dot{x}(t) = A\ x(t) + B\ u(t) \tag{7-2}$$

where $x(0) = x_0$ given and A and B as defined by (7-3) and (7-4), find the control law

$$u(t) = -F\ x(t) \tag{7-7}$$

which minimizes

$$J = \lim_{t_1 \to \infty} \left\{ \int_0^{t_1} \left[x'(t)\ Q\ x(t) + u'(t)\ R\ u(t) \right] dt \right\} \tag{7-8}$$

where Q and R are 4×4 and 2×2 identity matrices, respectively, and $u(t)$ is restricted to be piecewise constant over uniform time intervals of measure $\Delta = 0.5$ second. Under these restrictions, the dynamical equations and performance index J become

$$x\left[(i+1)\Delta\right] = \tilde{A}\ x(i\Delta) + \tilde{B}\ u(i\Delta) \tag{7-9}$$

and

$$J = \lim_{N \to \infty} \left\{ \sum_{i=0}^{N} \left[x'(i\Delta)\ \tilde{Q}\ x(i\Delta) + x'(i\Delta)\ \tilde{W}\ u(i\Delta) + u'(i\Delta)\ \tilde{R}\ u(i\Delta) \right] \right\} \tag{7-10}$$

199

where

$$\tilde{A} = e^{A\tau} \tag{7-11}$$

$$\tilde{B} = \int_0^\Delta e^{A\tau} B \, d\tau \tag{7-12}$$

$$\tilde{Q} = \int_0^\Delta e^{A'\tau} Q \, e^{A\tau} \, d\tau \tag{7-13}$$

$$\tilde{W} = 2 \int_0^\Delta e^{A'\tau} Q \, H(\tau,0) \, d\tau \tag{7-14}$$

$$\tilde{R} = \int_0^\Delta \left[R + H'(\tau,0) \, Q \, H(\tau,0) \right] d\tau \tag{7-15}$$

and

$$H(t,0) = \int_0^t e^{A\tau} B \, d\tau \tag{7-16}$$

Performing the control variable transformation,

$$u(i\Delta) = -\tilde{F} \, x(i\Delta) + v(i\Delta) \tag{7-17}$$

with

$$\tilde{F} = R^{-1} \frac{\tilde{W}'}{2} \tag{7-18}$$

leads to

$$x\left[(i+1)\Delta\right] = \hat{A} \, x(i\Delta) + \hat{B} \, v(i\Delta) \tag{7-19}$$

$$J = \lim_{N \to \infty} \left\{ \sum_{i=0}^N \left[x'(i\Delta) \, \hat{Q} \, x(i\Delta) + v'(i\Delta) \, \hat{R} \, v(i\Delta) \right] \right\} \tag{7-20}$$

where

$$\hat{A} = \tilde{A} - \tilde{B}\tilde{F} \tag{7-21}$$

$$\hat{B} = \tilde{B} \tag{7-22}$$

$$\hat{Q} = \tilde{Q} - \frac{\widetilde{W}\widetilde{F}}{2} \tag{7-23}$$

$$\hat{R} = \tilde{R} \tag{7-24}$$

Ignoring the initial value $x(0)$ in J gives the problem of minimizing with respect to v

$$\hat{J} = \lim_{N \to \infty} \left(\sum_{i=0}^{N} \left\{ x'\left[(i+1)\Delta\right] \hat{Q} \, x\left[(i+1)\Delta\right] + v'(i\Delta) \, \hat{R} \, v(i\Delta) \right\} \right) \tag{7-25}$$

subject to

$$x\left[(i+1)\Delta\right] = \hat{A} \, x(i\Delta) + \hat{B} \, v(i\Delta) \tag{7-19}$$

whose solution, if it exists, is given by

$$v(i\Delta) = -\hat{F} \, x(i\Delta) \tag{7-26}$$

The gain matrix for the original problem is then

$$F = \tilde{F} + \hat{F} \tag{7-27}$$

The complete problem can be solved by appropriately combining the sub-routines EXPINT (for $\tilde{A}$ and $\tilde{B}$), SAMPL (for $\tilde{Q}$, $\tilde{W}$, and $\tilde{R}$), PREFIL (for $\hat{A}$, $\hat{Q}$, and $\tilde{F}$), and ASYMREG (for $\hat{F}$). An executive program for constructing the solution follows. Data are input by using subroutine READ rather than being defined internally to the source program. The dimension of DUMMY is chosen to be the maximum of the requirements of each operation employing this storage vector. At the stage where matrices F and $\hat{F}$ are printed, the statement "CALL LNCNT (100)" is used to shift the printing to the top of the next page. Output data are presented after the executive program.

```
PROGRAM SAMDAT      74/74      OPT=1                    FTN 4.6+452                    79/04/13.  14.52.16                    PAGE    1


              PROGRAM SAMDAT(INPUT,OUTPUT,TAPE5=INPUT,TAPE6=OUTPUT)

        *****************************************************************************
        *
  5     *
        *          EXECUTIVE PROGRAM FOR SAMPLED-DATA REGULATOR PROBLEM
        *
        *
        *****************************************************************************
 10

              DIMENSION A(16),B(8),Q(16),R(4),ATIL(16),BTIL(8),QTIL(16),MTIL(8),
             1RTIL(4),FTIL(8),AHAT(16),QHAT(16),FHAT(8),DUMMY(116),P(16),F(8)
              DIMENSION NA(2),NB(2),NQ(2),NR(2),NATIL(2),NBTIL(2),NQTIL(2),NMTIL
 15          1(2),NRTIL(2),NFTIL(2),NAHAT(2),NQHAT(2),NFHAT(2),IOP(5),NP(2),NF(2
             2),NDUM(2)
              LOGICAL IDENT,DISC,NEWT,STABLE,FNULL
        C
        C     INPUT HOLLERITH DATA FOR TITLE OF OUTPUT
 20           CALL RDTITL
        C
        C     INPUT COEFFICIENT AND WEIGHTING MATRICES FOR CONTINUOUS SYSTEM
        C     CALL READ(4,A,NA,B,NB,Q,NQ,R,NR)
        C
 25     C     GENERATE EXP (A*DELT) AND (INTEGRAL EXP(A*TAU))*B
              DELT=.05
              IOP(1)=0
              N1 = (NA(1)**2)+1
        C
 30           CALL EXPINT(A,NA,ATIL,NATIL,DUMMY,NDUM,DELT,IOP,DUMMY(N1))
              CALL MULT(DUMMY,NDUM,B,NB,BTIL,NBTIL)
              CALL PRNT(ATIL,NATIL,4HATIL,1)
              CALL PRNT(BTIL,NBTIL,4HBTIL,1)
        C
 35     C     GENERATE DIGITAL PERFORMANCE INDEX WEIGHTING MATRICES
              IOP(2)=1
              CALL EQUATE(Q,NQ,QTIL,NQTIL)
              CALL EQUATE(R,NR,RTIL,NRTIL)
              CALL SAMPLA(A,NA,B,NB,QTIL,NQTIL,RTIL,NRTIL,MTIL,NMTIL,DELT,IOP,DUM
 40          1MY)
              CALL PRNT(QTIL,NQTIL,4HQTIL,1)
```

```
          CALL PRNT(WTIL,NWTIL,4HWTIL,1)
45        CALL PRNT(RTIL,NRTIL,4HRTIL,1)

C
C         FIND PREFILTER GAIN WHICH ELIMINATES CROSS-PRODUCT TERM
C         IN DIGITAL PERFORMANCE INDEX
          IOP(3)=1
50        CALL EQUATE(ATIL,NATIL,AHAT,NAHAT)
          CALL EQUATE(QTIL,NQTIL,QHAT,NQHAT)
          CALL PREFIL(AHAT,NAHAT,BTIL,NBTIL,QHAT,NQHAT,WTIL,NWTIL,RTIL,NRTIL
         1,FTIL,NFTIL,IOP,DUMMY)
          CALL PRNT(AHAT,NAHAT,4HAHAT,1)
          CALL PRNT(QHAT,NQHAT,4HQHAT,1)
55        CALL PRNT(FTIL,NFTIL,4HFTIL,1)

C
C         SOLVE FOR F HAT
          IDENT= .TRUE.
          DISC = .TRUE.
60        NEWT = .TRUE.
          STABLE = .FALSE.
          FNULL = .TRUE.
          ALPHA = .9
          IOP(1) = 1
65        IOP(2) = 0
          IOP(3) = 0
          IOP(4) = 1
          IOP(5) = 1
          CALL ASYMREG(AHAT,NAHAT,BTIL,NBTIL,H,NH,QHAT,NQHAT,RTIL,NRTIL,FHAT
         1,NFHAT,P,NP,IDENT,DISC,NEWT,STABLE,FNULL,ALPHA,IOP,DUMMY)
70        CALL ADD(FTIL,NFTIL,FHAT,NFHAT,F,NF)

C
C         OUTPUT F AND F HAT GAINS
          CALL LNCNT(100)
          PRINT 100
75  100   FORMAT(*          FOR THE ORIGINAL SAMPLEC-DATA PROBLEM**)
          CALL PRNT(FHAT,NFHAT,4HFHAT,1)
          CALL PRNT(F,NF,4HF  ,1)

C
          STOP
80        END
```

203

C/ Output from ORACLS

OUTPUT FOR SAMPLED-DATA REGULATOR PROBLEM

```
   A    MATRIX        4 ROWS           4 COLUMNS
-2.6000000E+00    2.5000000E-01   -3.8000000E+01    0.
-7.5000000E-02   -2.7000000E-01    4.4000000E+00    0.
 7.8000000E-02   -9.9000000E-01   -2.3000000E-01    5.2000000E-02
 1.0000000E+00    7.8000000E-02    0.               0.
```

```
   B    MATRIX        4 ROWS           2 COLUMNS
 1.7000000E+01    7.0000000E+00
 8.2000000E-01   -3.2000000E+00
 0.               4.6000000E-02
 0.               0.
```

```
   Q    MATRIX        4 ROWS           4 COLUMNS
 1.0000000E+00    0.               0.               0.
 0.               1.0000000E+00    0.               0.
 0.               0.               1.0000000E+00    0.
 0.               0.               0.               1.0000000E+00
```

```
   R    MATRIX        2 ROWS           2 COLUMNS
 1.0000000E+00    0.
 0.               1.0000000E+00
```

```
  ATIL MATRIX        4 ROWS           4 COLUMNS
 8.7461216E-01    5.6206553E-02   -1.7644035E+00   -2.3524288E-03
-3.0698872E-03    9.8114690E-01    2.1998505E-01    2.8616468E-04
 3.7743137E-03   -4.8707923E-02    9.7958437E-01    2.5728852E-03
 4.6820759E-02    4.9177112E-03   -4.4489768E-02    9.9996068E-01
```

```
  BTIL MATRIX        4 ROWS           2 COLUMNS
 7.9692392E-01    3.2234606E-01
 3.9250218E-02   -1.5895733E-01
 6.1800654E-04    6.8676137E-03
 2.0435169E-02    7.9852845E-03
```

```
  QTIL MATRIX        4 ROWS           4 COLUMNS
 4.3959404E-02    8.7398740E-04   -4.1390797E-02    1.1608794E-03
 8.7398740E-04    4.9227561E-02    2.8619327E-03    1.1266907E-04
-4.1390797E-03    2.8619327E-03    1.0367223E-01   -6.3835981E-04
 1.1608794E-03    1.1266907E-04   -6.3835981E-04    4.9991178E-02
```

```
  MTIL MATRIX        4 ROWS           2 COLUMNS
```

OUTPUT FOR SAMPLED-DATA REGULATOR PROBLEM

3.7319029E-02   1.5229661E-02
3.1985357E-03 -7.3838878E-03
-4.8263666E-02 -2.0667855E-02
6.4086387E-04   2.4991144E-04

RTIL MATRIX   2 ROWS   2 COLUMNS
6.0964766E-02   4.3451993E-03
4.3451993E-03   5.2234106E-02

AHAT MATRIX   4 ROWS   4 COLUMNS
5.9854577E-01   5.4764994E-02 -1.4049026E+00 -7.0631127E-03
4.4954715E-03   9.6826103E-01   2.0868737E-01   3.9798041E-04
2.7592406E-03 -4.8223984E-02   9.8096098E-01   2.5606174E-03
3.9775917E-02   4.8601865E-03 -3.5637777E-02   9.9984043E-01

QHAT MATRIX   4 ROWS   4 COLUMNS
3.7487559E-02   8.4516580E-04 -3.2962128E-02   1.0504445E-03
8.4516580E-04   4.8906642E-02   2.8635932E-03   1.1174823E-04
-3.2962128E-02   2.8635932E-03   9.2886064E-02 -4.9458179E-04
1.0504445E-03   1.1174823E-04 -4.9458179E-04   4.9997293E-02

FTIL MATRIX   2 ROWS   4 COLUMNS
2.9744352E-01   3.1456861E-02 -3.8400852E-01   5.1158474E-03
1.2103929E-01 -7.3297516E-02 -1.6589418E-01   1.9666527E-03

COMPUTATION OF F SUCH THAT A-BF IS ASYMPTOTICALLY STABLE IN THE DISCRETE SENSE

A   MATRIX   4 ROWS   4 COLUMNS
6.6505086E-01   6.0849993E-02 -1.5610029E+00 -7.8481252E-03
4.9945683E-03   1.0758456E+00   2.3187486E-01   4.4220046E-04
3.0658229E-03 -5.3582205E-02   1.0899566E+00   2.8451304E-03
4.4195464E-02   5.4002072E-03 -3.9597530E-02   1.1109338E+00

B   MATRIX   4 ROWS   2 COLUMNS
8.8547102E-01   3.5816229E-01
4.3611353E-02 -1.7661925E-01
6.8667393E-04   7.6306819E-03
2.2706743E-02   8.8725383E-03

OUTPUT FOR SAMPLED-DATA REGULATOR PROBLEM

ALPHA =   .33873839E+00

F   MATRIX   2 ROWS   4 COLUMNS
8.1498121E-01   2.2921930E+00 -3.0628590E+01   1.4899924E+01
1.0955671E-01 -5.7422787E+00   8.0373720E+01 -3.4265557E+00

A-BF MATRIX   4 ROWS   4 COLUMNS
-9.5830473E-02   8.7847250E-02 -3.2271096E+00 -1.1974036E+01
-1.1197641E-02 -3.8317030E-02   1.5763176E+01 -1.2545593E+00
 1.6702042E-03 -1.1338652E-02   4.9768221E-01   1.8760698E-02
 2.4716663E-02   4.3028501E-03 -5.7271536E-02   8.0302221E-01

EIGENVALUES OF A-BF

EIGN MATRIX   4 ROWS   2 COLUMNS
2.3419229E-01   3.3528007E-01
2.3419229E-01 -3.3528007E-01
3.4908617E-01   3.1664598E-01
3.4908617E-01 -3.1664598E-01

MODULI CF EIGENVALUES OF A-BF

MOC MATRIX   4 ROWS   1 COLUMNS
4.0897281E-01
4.0897281E-01
4.7130228E-01
4.7130228E-01

PROGRAM TO SOLVE DISCRETE STEADY-STATE RICCATI EQUATION BY THE NEWTON ALGORITHM

A   MATRIX   4 ROWS   4 COLUMNS
5.9854577E-01   5.4764994E-02 -1.4049026E+00 -7.0633127E-03
4.4954715E-03   9.6826103E-01   2.0868737E-01   3.9798041E-04
2.7592406E-03 -4.8239984E-02   9.8096098E-01   2.5606174E-03

206

OUTPUT FOR SAMPLED-DATA REGULATOR PROBLEM

3.9775917E-02   4.8601865E-03   -3.5637777E-02   9.9840043E-01

B MATRIX   4 ROWS   2 COLUMNS
```
7.9692392E-01   3.2234606E-01
3.9250218E-02  -1.5895733E-01
6.1800654E-04   6.8676137E-03
2.0435169E-02   7.9852845E-03
```

Q MATRIX   4 ROWS   4 COLUMNS
```
3.7687559E-02   8.4516580E-04  -3.2962128E-02   1.0504445E-03
8.4516580E-04   4.8906642E-02   2.8635932E-04   1.1174823E-04
-3.2962128E-02  2.8635932E-03   9.2886064E-02  -4.9481796E-04
1.0504445E-03   1.1174823E-04  -4.9458179E-04   4.9972993E-02
```

H IS AN IDENTITY MATRIX

R MATRIX   2 ROWS   2 COLUMNS
```
6.0964766E-02   4.3451993E-03
4.3451993E-03   5.2234106E-02
```

INITIAL F MATRIX

F MATRIX   2 ROWS   4 COLUMNS
```
8.1498121E-01   2.2921930E+00  -3.0628590E+01   1.4899924E+01
1.0955671E-01  -5.7422787E+00   8.0373720E+01  -3.4265557E+00
```

FINAL VALUES OF P AND F AFTER   10 ITERATIONS TO CONVERGE

P MATRIX   4 ROWS   4 COLUMNS
```
5.1520273E-02   1.0590966E-02  -9.5089721E-02   5.6173730E-02
1.0590966E-02   3.6895816E-01  -3.4132485E-01   4.2819085E-02
-9.5089721E-02 -3.4132485E-01   2.9211059E+00  -1.3498471E-01
5.6173730E-02   4.2819085E-02  -1.3498471E-01   1.0635575E+00
```

F MATRIX   2 ROWS   4 COLUMNS
```
2.7657460E-01   4.4589747E-01  -1.5678530E+00   6.7515437E-01
8.2414674E-02  -9.4596876E-01   4.6313338E-01   1.3877541E-01
```

207

OUTPUT FOR SAMPLED-DATA REGULATOR PROBLEM

RESIDUAL ERROR IN RICCATI EQUATION

```
    EROR MATRIX      4 ROWS          4 COLUMNS
-3.1086245E-15  -4.2184475E-15   1.1990409E-14  -4.6185278E-14
-4.2743586E-15  -7.1054274E-15   1.7763568E-14  -7.5495166E-14
 1.1990409E-14   1.7763568E-14  -4.2632564E-14   1.9895197E-13
-4.6851412E-14  -7.5051076E-14   1.9895197E-13  -8.3133500E-13
```

EIGENVALUES OF P

```
    EVLP MATRIX      4 ROWS          1 COLUMNS
4.5717003E-02
3.2330141E-01
1.0566117E+00
2.9795117E+00
```

CLOSED-LOOP RESPONSE MATRIX A-BF

```
    A-BF MATRIX      4 ROWS          4 COLUMNS
3.5157081E-01   4.3479404E-03  -3.0473224E-01  -5.8884369E-01
6.7402744E-03   8.0039079E-01   3.4384439E-01  -4.0426071E-03
2.0223236E-03  -4.2003004E-02   9.7874931E-01   1.1903116E-03
3.3465964E-02   3.3020260E-03  -7.2966874E-03   9.8493537E-01
```

EIGENVALUES OF A-BF

```
3.8573552E-01   0.
8.8933093E-01   7.9976185E-02
8.8933093E-01  -7.9976185E-02
9.5124891E-01   0.
```

FCR THE ORIGINAL SAMPLED-DATA PROBLEM

```
    FHAT MATRIX      2 ROWS          4 COLUMNS
2.7657460E-01   4.4589747E-01  -1.5678530E+00   6.7515437E-01
8.2414674E-02  -9.4596876E-01   4.6313338E-01   1.3877541E-01
```

```
    F MATRIX         2 ROWS          4 COLUMNS
5.7401812E-01   4.7735433E-01  -1.9518615E+00   6.8027022E-01
2.0345396E-01  -1.0192663E+00   2.9723921E-01   1.4074207E-01
```

IV/ MODEL FOLLOWING

A/ Problem Statement

This problem demonstrates a model-following calculation and illustrates the output format of the transient response subroutine TRANSIT.

Given the linear system

$$\dot{x}(t) = A\,x(t) + B\,u(t) \tag{7-2}$$

with $x(0) = x_0$ given and A and B as defined by (7-3) and (7-4), use the explicit model-following procedure to find a control law which causes the state $x(t)$ to track the state $\tilde{x}_m(t)$ of the system,

$$\dot{\tilde{x}}_m(t) = \tilde{A}_m\,\tilde{x}_m(t) + \tilde{B}_m\,\tilde{u}_m(t) \tag{7-28}$$

where $\tilde{x}_m(0) = \tilde{x}_m^0$ is given and the step input is

$$\tilde{u}_m(t) = \begin{bmatrix} 2 \\ 0 \end{bmatrix} \tag{7-29}$$

with

$$\tilde{A}_m = \begin{bmatrix} -0.981 & 0.177 & -10.0 & 0.0 \\ 0.030 & -0.092 & 5.23 & 0.0 \\ 0.0 & -1.0 & -0.732 & 0.052 \\ 1.0 & 0.0 & 0.0 & 0.0 \end{bmatrix} \tag{7-30}$$

$$\tilde{B}_m = \begin{bmatrix} 6.34 & 4.58 \\ -0.690 & -2.65 \\ 0.0 & 0.0 \\ 0.0 & 0.0 \end{bmatrix} \tag{7-31}$$

Generally, the state and model dynamics represent two sets of linearized equations and $u(t)$ is to cause the system represented by $x(t)$ to perform like the model $\tilde{x}_m(t)$ when the model is subjected to a particular step input. Data for the model dynamics are from handling-quality model 3 of [7-2].

The solution to the posed problem can be found by using the subroutine EXPMDFL of ORACLS. The problem can be stated in the format of EXPMDFL by redefining the model dynamics as

$$\dot{x}_m(t) = A_m\, x_m(t) \tag{7-32}$$

with

$$x_m(0) = \begin{bmatrix} \tilde{x}_m^0 \\ 2 \\ 0 \end{bmatrix} \tag{7-33}$$

and

$$y_m(t) = H_m\, x_m(t) \tag{7-34}$$

where

$$x_m(t) = \begin{bmatrix} \tilde{x}_m(t) \\ \tilde{u}_m(t) \end{bmatrix} \tag{7-35}$$

$$A_m = \begin{bmatrix} \tilde{A}_m & \vdots & \tilde{B}_m \\ \text{------} & \vdots & \text{------} \\ 0_{2\times4} & \vdots & 0_{2\times2} \end{bmatrix} \tag{7-36}$$

210

and

$$H_m = \begin{bmatrix} I_4 & \vdots & 0_{4\times2} \end{bmatrix}$$ (7-37)

The matrix I_4 designates a 4×4 identity matrix and $0_{4\times2}$ a 4×2 null matrix. Weighting matrices in EXPMDFL are chosen to be

$$Q = \text{diag} (10.0, 10.0, 10.0, 10.0),$$ (7-38)

and R is the 2×2 identity matrix. A 26-second transient response is evaluated using TRANSIT with

$$x(0) = \tilde{x}^o_m = 0$$ (7-39)

The executive program and output data follow.

B/ Executive Program

```
           PROGRAM MODFOL(INPUT,OUTPUT,TAPE5=INPUT,TAPE6=OUTPUT)

     ************************************************************
     *                                                          *
     *     EXECUTIVE PROGRAM FOR MODEL FOLLOWING PROBLEM        *
     *                                                          *
     ************************************************************

           DIMENSION A(16),B(8),AM(36),HM(24),Q(16),R(4),F(20),P(40),DUMMY(70
          10),AC(100),BC(20),X(10)
           DIMENSION NA(2),NB(2),NAM(2),NHM(2),NQ(2),NR(2),NF(2),NP(2),NAC(2)
          1,NBC(2),IOP(5),NDUM(2),T(2),NX(2)
           LOGICAL DISC,NEWT,STABLE,HIDENT,HMIDENT
     C
     C     INPUT HOLLERITH DATA FOR TITLE OF OUTPUT
           CALL RDTITL
     C
     C     INPUT COEFFICIENT MATRICES FOR PLANT AND MODEL
           CALL READ(5,A,NA,B,NB,AM,NAM,HM,NHM,X,NX)
     C
     C     DEFINE WEIGHTING MATRICES
           NQ(1)= NA(1)
           NQ(2)= NQ(1)
           NR(1)= NB(2)
           NR(2)= NR(1)
           CALL UNITY(Q,NQ)
           CALL SCALE(Q,NQ,Q,NQ,10.)
           CALL UNITY(R,NR)
     C
     C     COMPUTE MODEL-FOLLOWING GAINS
           IOP(1) = 1
           IOP(2) = 1
           IOP(3) = 0
           IOP(4) = 0
           IOP(5) = 1
           DISC = .FALSE.
           NEWT = .FALSE.
           STABLE = .FALSE.
```

```
        HIDENT=.TRUE.
        HMIDENT = .FALSE.

C       CALL EXPMDFL(A,NA,B,NB,H,NH,AM,NAM,HM,NHM,Q,NQ,R,NR,F,NF,P,NP,HIDE
       1NT,HMIDENT,DISC,NENT,STABLE,FNULL,ALPHA,IOP,DUMMY)

C
C       GENERATE MATRICES FOR TRANSIENT RESPONSE SUBROUTINE
        NDUM(1) = NAM(1)
        NDUM(2) = NA(1)
        CALL NULL(DUMMY,NDUM)
        CALL JUXTR(A,NA,DUMMY,NDUM,AC,NAC)
        NDUM(1) = NA(1)
        NDUM(2) = NAM(1)
        CALL NULL(DUMMY,NDUM)
        N1 = NDUM(1)*NDUM(2) + 1
        CALL JUXTR(DUMMY,NDUM,AM,NAM,DUMMY(N1),NHM)
        N2=NHM(1)*NHM(2)+N1
        CALL JUXTC(AC,NAC,DUMMY(N1),NHM,DUMMY(N2),NDUM)
        CALL EQUATE(DUMMY(N2),NDUM,AC,NAC)
        NDUM(1) = NAM(1)
        NDUM(2) = NB(2)
        CALL NULL(DUMMY,NDUM)
        CALL JUXTR(B,NB,DUMMY,NDUM,BC,NBC)

C
C       COMPUTE TRANSIENT RESPONSE TO OBSERVE MODEL-FOLLOWING ACCURACY
        IOP(1) = 0
        IOP(2) = 0
        IOP(3) = 0
        T(1) = 26.
        T(2) = 2.

C       CALL LNCNT(100)
        PRINT 100,NA(1),NAM(1)
100     FORMAT(*    IN THE TRAJECTORY OUTPUT TO FOLLOW THE FIRST*I4* CO
       1LUMNS CORRESPOND TO X TRANSPOSE*/*    AND THE NEXT*I4* COLUMNS TO X
       2M TRANSPOSE*)

C       CALL TRANSIT(AC,NAC,BC,NBC,H,NH,G,NG,F,NF,V,NV,T,X,NX,DISC,STABLE,
       1IOP,DUMMY)

C       STOP
        END
```

C/ Output from ORACLS

OUTPUT FOR MODEL FOLLOWING PROBLEM

```
 A   MATRIX      4 ROWS      4 COLUMNS
-2.6000000E+00   2.5000000E-01  -3.8000000E+01   0.
-7.5000000E-02  -2.7000000E-01   4.4000000E+00   0.
 7.8000000E-02  -9.9000000E-01  -2.3000000E-01   5.2000000E-02
 1.0000000E+00   7.8000000E-02   0.              0.
```

```
 B   MATRIX      4 ROWS      2 COLUMNS
 1.7000000E+01   7.0000000E+00
 8.2000000E-01  -3.2000000E+00
 0.              4.6000000E-02
 0.              0.
```

```
 AM  MATRIX      6 ROWS      6 COLUMNS
-9.8100000E-01   1.7700000E-01  -1.0000000E+01   0.              6.3400000E+00   4.5800000E+00
 3.0000000E-02  -9.2000000E-02   5.2300000E+00   0.             -6.9000000E-01  -2.6500000E+00
-1.0000000E+00  -1.0000000E+00  -7.3200000E-01   5.2000000E-02   0.              0.
 1.0000000E+00   0.              0.              0.              0.              0.
 0.              0.              0.              0.              0.              0.
 0.              1.0000000E+00   0.              1.0000000E+00   0.              0.
```

```
 HM  MATRIX      4 ROWS      6 COLUMNS
 1.0000000E+00   0.              0.              0.              0.              0.
 0.              1.0000000E+00   0.              0.              0.              0.
 0.              0.              1.0000000E+00   0.              0.              0.
 0.              0.              0.              1.0000000E+00   0.              0.
```

```
 X   MATRIX     10 ROWS      1 COLUMNS
 0.
 0.
 0.
 0.
 0.
 0.
 0.
 2.0000000E+00
 0.
```

PROGRAM TO SOLVE ASYMPTOTIC CONTINUOUS EXPLICIT MODEL-FOLLOWING PROBLEM

PLANT DYNAMICS

```
    A   MATRIX      4 ROWS           4 COLUMNS
-2.6000000E+00   2.5000000E-01  -3.8000000E+01   0.
-7.5000000E-02  -2.7000000E-01   4.4000000E+00   0.
 7.8000000E-02  -9.9000000E-01  -2.3000000E-01   5.2000000E-02
 1.0000000E+00   7.8000000E-02   0.              0.
```

```
    B   MATRIX      4 ROWS           2 COLUMNS
 1.7000000E+01   7.0000000E+00
 8.2000000E-01  -3.2000000E-01
 0.              4.6000000E-02
 0.              0.
```

H IS AN IDENTITY MATRIX

MODEL DYNAMICS

```
    AM  MATRIX      6 ROWS           6 COLUMNS
-9.8100000E-01   1.7700000E-01  -1.0000000E+00   0.   6.3400000E+00   4.5800000E+00
 3.0000000E+00  -9.2000000E-02   5.2300000E+00   0.  -6.9000000E-01  -2.6500000E+00
 0.             -1.0000000E+00  -7.3200000E-01   5.2300000E-02   0.   0.
 1.0000000E+00   0.              0.              0.   0.   0.
 0.              0.              0.              0.   0.   0.
 0.              0.              0.              0.   0.   0.
```

```
    HM  MATRIX      4 ROWS           6 COLUMNS
 1.0000000E+00   0.              0.              0.   0.   0.
 0.              1.0000000E+00   0.              0.   0.   0.
 0.              0.              1.0000000E+00   0.   0.   0.
 0.              0.              0.              1.0000000E+00   0.   0.
```

215

OUTPUT FOR MODEL FOLLOWING PROBLEM

WEIGHTING MATRICES

```
      Q   MATRIX        4 ROWS            4 COLUMNS
1.0000000E+01   0.             0.             0.
0.              1.0000000E+01  0.             0.
0.              0.             1.0000000E+01  0.
0.              0.             0.             1.0000000E+01

      R   MATRIX        2 ROWS            2 COLUMNS
1.0000000E+00   0.
0.              1.0000000E+00
```

CONTROL LAW U = -F(COL.(X,XM)),   F = (F11,F12)

PART OF F MULTIPLYING X

```
   F11  MATRIX        2 ROWS            4 COLUMNS
2.8715216E+00   -1.1867788E+00  -2.3439987E+00   3.0218811E+00
1.0990931E+00   -3.0362141E+00   1.5921904E+00   9.3351305E-01

   P11  MATRIX        4 ROWS            4 COLUMNS
1.6784659E-01    2.2109123E-02  -1.1051966E-01   1.7350075E-01
2.2109123E-02    9.8893140E-01  -5.7385910E-01   8.8254034E-02
-1.1051966E-01  -5.7385910E-01   1.1510412E+01   3.0884697E-02
1.7350075E-01    8.8254034E-02   3.0884697E-02   1.0164474E+01
```

EIGENVALUES OF P11

```
1.6346846E-01
9.5710942E-01
1.0167942E+01
1.1543144E+01
```

PLANT CLOSED-LOOP RESPONSE MATRIX A - BF11

OUTPUT FOR MODEL FOLLOWING PROBLEM

```
-5.9109518E+01    1.3282587E+00   -9.2055556E+00   -5.7906570E+01
 1.0874501E+00   -1.0959044E+01    1.1421516E+01    5.0929927E-01
 2.7441719E-02   -8.5033415E-01   -3.0324076E-01    9.0583996E-03
 1.0000000E+00    7.8000000E-02    0.               0.
```

EIGENVALUES OF CLOSED-LOOP RESPONSE MATRIX

```
-1.0076624E+00    0.
-1.2943964E+00    0.
-9.9317854E+00    0.
-5.8137959E+01    0.
```

PART OF F MULTIPLYING XM

```
 F12 MATRIX    2 ROWS         6 COLUMNS
-2.9990036E+00  -1.0535303E+00   9.7142903E-01  -3.0178671E+00  -5.5004933E-01  -4.3236218E-01
-9.5061040E-01   2.9377885E+00  -1.5695172E+00  -9.2056907E-01   8.1556250E-01   4.4887921E-01

 P12 MATRIX    4 ROWS         6 COLUMNS
-1.7290904E-01  -1.6058975E-02   3.6971673E-02  -1.7312936E-01  -2.0297024E-02  -1.9460473E-02
-7.2621845E-02  -9.5186308E-01   4.1818364E-01  -9.1058595E-02  -2.4999989E-01  -1.2382211E-01
 5.9484706E-01   9.2161895E-02  -1.0655026E+01  -1.1101935E-03   3.4270003E+00   4.1059083E+00
-1.5393360E-01   1.4943541E-01   4.0497750E-01  -1.0136463E+01   6.3234900E-02  -3.7895492E-01
```

217

IN THE TRAJECTORY OUTPUT TO FOLLOW THE FIRST 4 COLUMNS CORRESPOND TO X TRANSPOSE
AND THE NEXT 6 COLUMNS TO XM TRANSPOSE

COMPUTATION OF TRANSIENT RESPONSE FOR THE CONTINUOUS   SYSTEM

```
A    MATRIX      10 ROWS         10 COLUMNS
-2.6000000E+00   2.5000000E-01  -3.8000000E+01   0.             0.            -9.8100000E-01   1.7700000E-01  -1.0000000E+01   0.
 0.
-7.5000000E-02  -2.7000000E-01   4.4000000E+00   0.             0.             3.0000000E-02  -9.2000000E-02   5.2300000E+00   0.
 0.
 7.8000000E-02  -9.9000000E-01  -2.3000000E-01   5.2000000E-02  0.            -1.0000000E+00  -1.0000000E+00  -7.3200000E-01   0.
 0.
 1.0000000E+00   7.8000000E-02   0.              0.             1.0000000E+00  0.              0.              0.              0.
 0.
 0.              0.              6.3400000E+00   4.5800000E+00  0.             1.0000000E+00   0.              0.              0.
 0.
 0.              0.             -6.9000000E-01  -2.6500000E+00  0.             0.              0.              0.              0.
 0.
 5.2000000E-02   0.              0.              0.             0.             0.              0.              0.              0.
 0.
 0.              0.              0.              0.             0.             0.              0.              0.              0.
 0.
 0.              0.              0.              0.             0.             0.              0.              0.              0.
 0.
 0.              0.              0.              0.             0.             0.              0.              0.              0.
 0.
```

```
B    MATRIX      10 ROWS          2 COLUMNS
 1.7000000E+01   7.0000000E+00
 8.2000000E-01  -3.2000000E+00
 0.              3.2000000E+00
 0.              4.6000000E-02
 0.              0.
 0.              0.
 0.              0.
 0.              0.
 0.              0.
 0.              0.
```

H IS A NULL MATRIX

OUTPUT FOR MODEL FOLLOWING PROBLEM

G IS A NULL MATRIX

```
     F    MATRIX        2 ROWS        10 COLUMNS
 2.8712216E+00  1.1867788E+00 -2.3493987E+00  3.0218811E+00 -2.9990036E+00 -1.0535303E+00  9.7142903E-01
-3.0178671E+00 -5.5004933E-01 -4.3236218E-01
 1.0990931E+00 -3.0362141E+00  1.5921904E+00  9.3351305E-01 -9.5061040E-01  2.9377885E+00 -1.5695172E+00
-9.2056907E-01  8.1562650E-01  4.4887921E-01
```

V IS A NULL MATRIX

COMPUTATION OF THE MATRIX EXPONENTIAL EXP(A T) BY THE SERIES METHOD

```
     A    MATRIX       10 ROWS       10 COLUMNS
-5.9105518E+01  1.3282587E+00 -9.2055556E+00 -5.7906570E+01  5.7637333E+01 -2.6545044E+00 -5.5276735E+00
 5.7747725E+01  3.6419010E+00  4.2080025E+00
 1.0874501E+00 -1.0959044E+01  1.1421516E+01  5.0929927E-01 -5.8277036E-01  1.0264818E+01 -5.8190267E+00
-4.7116998E-01  3.0608404E+00  1.7909505E+00
 2.7441719E-02 -8.5033415E-01 -3.0324076E-01  9.0583996E-03  4.3728078E-02 -1.3513327E-01  7.2197769E-02
 4.2346177E-02 -3.7515875E-02 -2.0648444E-02
 1.0000000E+00  7.8000000E-02  0.           0.           0.           0.           0.
 0.             0.             0.
 0.             0.             6.3400000E+00  4.5800000E+00 -9.8100000E-01  1.7700000E-01 -1.0000000E+01
 0.             0.             0.
 0.             0.            -6.9000000E-01 -2.6500000E+00  3.0000000E+00 -9.2000000E-02  5.2300000E+00
 0.             0.             0.
 0.             5.2000000E-02  0.           0.           0.          -1.0000000E+00 -7.3200000E-01
 0.             0.             0.
 5.2000000E-02  0.             0.           0.           1.0000000E+00  0.           0.
 0.             0.             0.
 0.             0.             0.           0.           0.           0.           0.
 0.             0.             0.
 0.             0.             0.           0.           0.           0.           0.
 0.             0.             0.
```

T =   .20000000E+01

219

OUTPUT FOR MODEL FOLLOWING PROBLEM

```
     EXFA MATRIX      10 ROWS      10 COLUMNS
-2.3891131E-03 -1.3251850E-03 -1.1385082E-04 -1.3738924E-01  1.1955605E-01  1.0625337E-01  2.1427525E+00
 1.2605585E-01  3.9547601E+00 -1.4409641E+00                  6.2644729E-02  1.4534787E-02 -1.0599349E+00
-1.3074054E-04 -8.5670545E-03  9.7002193E-02 -8.5196158E-04   9.2176534E-03  2.2259046E-01 -1.5515221E-01
 5.3247219E-02  7.2397266E-01  3.8807667E-01
 1.4763163E-05 -7.2188507E-03  8.2534175E-02  6.5207081E-03
-1.6411698E-02  8.2364845E-02  4.8831924E-01                  8.4821386E-01  2.0037391E+00 -1.2295579E-01
 2.3921962E-03  2.2911988E-03 -1.0898440E-02  1.3680976E-01
 7.6035997E-01  5.8562550E+00  2.3363158E-01                  1.2107298E-01  1.1670875E-01  2.1381511E+00
 0.             0.             0.                             -7.8748841E-03  4.0115443E+00 -1.4422067E+00  4.8006911E-02 -3.5454142E-02 -1.0445C38E+00
 0.             5.4105409E-02  3.0701214E-01  1.1717037E-01    1.2870845E-03  2.0066917E-01 -1.1693378E-02
 0.            -9.3881938E-03  1.7719426E-01  5.5703078E-01    8.5227516E-01  2.0172177E+00 -1.5144C68E-01
 0.             8.9663838E-01  5.8926293E+00  2.6304712E-01    0.             0.             0.
 0.             0.             1.0000000E+00  0.               0.             0.             0.
 0.             0.             0.             1.0000000E+00
```

STRUCTURE OF PRINTING TO FOLLOW

```
TIME CR STAGE
STATE   - X TRANSPOSE - FROM DX = AX + BU
OUTPUT  - Y TRANSPOSE - FROM Y  = HX + GU    IF DIFFERENT FROM X
CONTROL - U TRANSPOSE - FROM U  = -FX + V
```

```
0.
0.          0.          2.0000000E+00  0.          0.          0.
0.
0.          0.          2.0000000E+00  0.          0.          0.          0.
```

OUTPUT FOR MODEL FOLLOWING PROBLEM

1.1000987E+00 -1.6311250E+00

.2000000E+01

7.9095202E+00 1.4479453E+00 1.6472969E-01 1.1712510E+01 8.0230886E+00 6.1402428E-01 3.5438851E-01
1.1785259E+01 2.0000000E+00 0.

1.5927986E+00 1.0410762E-01

.4000000E+01

9.5489317E+00 2.2039396E+00 2.0659338E-01 3.0288443E+01 9.7310584E+00 1.2449044E+00 3.7314436E-01
3.0375205E+01 2.0000000E+00 0.

1.8227148E+00 1.0305996E-01

.6000000E+01

9.6466669E+00 3.2716314E+00 1.7391236E-01 4.9680598E+01 9.9051858E+00 2.2907575E+00 3.2719551E-01
4.9769107E+01 2.0000000E+00 0.

1.7942873E+00 6.0926728E-02

OUTPUT FOR MODEL FOLLOWING PROBLEM

.8000000E+01

9.4589500E+00 4.3502637E+00   2.1320376E-01   6.9319800E+01   9.7973579E+00   3.5575370E-01
6.9423499E+01 2.0000000E+00   0.

1.8504096E+00 3.1082298E-02

.1000000E+02

9.3974701E+00 5.3531655E+00   2.4666911E-01   8.9012688E+01   9.8153021E+00   3.8519443E-01
8.9105733E+01 2.0000000E+00   0.

1.9097142E+00 -9.6113310E-03

.1200000E+02

9.3439028E+00 6.3634006E+00   2.6358106E-01   1.0866891E+02   9.8410301E+00   5.3497710E+00   3.9885896E-01
1.0876250E+02 2.0000000E+00   0.

1.9444176E+00 -5.5910245E-02

.1400000E+02

222

OUTPUT FOR MODEL FOLLOWING PROBLEM

ORACLS PROGRAM

9.2585870E+00 7.3879771E+00 2.8394186E-01 1.2832999E+02 9.8352651E+00 6.3648203E+00 4.1484140E-01
1.2842441E+02 2.0000000E+00 0.

1.9814221E+00 -9.9857508E-02

.1600000E+02

9.1761171E+00 8.4088289E+00 3.0715116E-01 1.4799891E+02 9.8236999E+00 7.3756775E+00 4.3374640E-01
1.4809427E+02 2.0000000E+00 0.

2.0230646E+00 -1.4313242E-01

.1800000E+02

9.0997321E+00 9.4274267E+00 3.2918291E-01 1.6766858E+02 9.8355191E+00 8.3841989E+00 4.5170504E-01
1.6776481E+02 2.0000000E+00 0.

2.0634769E+00 -1.8702727E-01

.2000000E+02

9.0217633E+00 1.0447312E+01 3.5081025E-01 1.8733946E+02 9.8370988E+00 9.3941185E+00 4.6920740E-01
1.8743654E+02 2.0000000E+00 0.

223

OUTPUT FOR MODEL FOLLOWING PROBLEM

2.1031212E+00 -2.3097232E-01

.2200000E+02

8.9427645E+00 1.1467616E+01 3.7275076E-01 2.0701295E+02 9.8376666E+00 1.0404454E+01 4.8698248E-01
2.0711089E+02 2.0000000E+00 0.

2.1431502E+00 -2.7478692E-01

.2400000E+02

8.8642480E+00 1.2487737E+01 3.9472902E-01 2.2668872E+02 9.8387229E+00 1.1414584E+01 5.0481197E-01
2.2678753E+02 2.0000000E+00 0.

2.1832840E+00 -3.1861843E-01

.2600000E+02

8.7865678E+00 1.3507953E+01 4.1663924E-01 2.4636654E+02 9.8399126E+00 1.2424811E+01 5.2257858E-01
2.4646621E+02 2.0000000E+00 0.

2.2233270E+00 -3.6248104E-01

224

V/ KALMAN-BUCY FILTER

A/ Problem Statement

This problem illustrates the calculation of an asymptotic optimal estimator. Consider the linear system

$$\dot{x}(t) = A\ x(t) + B\ u(t) + \xi(t) \tag{7-40}$$

with A and B from (7-3) and (7-4) and $\xi(t)$ a zero-mean Gaussian
white-noise process with constant intensity $Q = Q' \geqq 0$. If the system
is digitally controlled using a zero-order hold with sampling times equal
to integral multiples $\Delta = 0.05$ second, the state equation becomes

$$x\big[(i+1)\Delta\big] = \tilde{A}\ x(i\Delta) + \tilde{B}\ u(i\Delta) + \tilde{\xi}(i\Delta) \qquad (i = 0,1,\ldots) \tag{7-41}$$

where

$$\tilde{A} = e^{A\Delta} \tag{7-42}$$

$$\tilde{B} = \int_0^\Delta e^{A\tau} B\ d\tau \tag{7-43}$$

and $\tilde{\xi}(i\Delta)$ is a zero-mean Gaussian discrete white-noise process with
variance matrix

$$\tilde{Q} = \int_0^\Delta e^{A\tau} Q e^{A'\tau}\ d\tau \tag{7-44}$$

Observations are made at the sampling instants and are modeled by

$$y(i\Delta) = H\ x(i\Delta) + D\ u(i\Delta) + \eta(i\Delta) \tag{7-45}$$

where

$$H = \begin{bmatrix} 1.0 & 0.0 & 0.0 & 0.0 \\ 0.0 & 1.0 & 0.0 & 0.0 \\ -0.0852 & -0.0421 & -2.24 & -0.0021 \end{bmatrix} \tag{7-46}$$

$$D = \begin{bmatrix} 0.0 & 0.0 \\ 0.0 & 0.0 \\ 0.713 & -0.403 \end{bmatrix} \tag{7-47}$$

and $\eta(i\Delta)$ is a zero-mean Gaussian discrete white-noise process with variance $R = R' > 0$. Assuming $\tilde{\xi}(i\Delta)$ and $\eta(i\Delta)$ are mutually uncorrelated, find a Kalman-Bucy predictor to asymptotically estimate the state $x(i\Delta)$ from knowledge of the measurements $y(i\Delta)$ and control inputs $u(i\Delta)$ up to time $(i-1)\Delta$.

If the estimate of x is denoted by $\hat{x}$, the predictor, [7-3], has the structure

$$\hat{x}\left[(i+1)\Delta\right] = \tilde{A}\ x(i\Delta) + \tilde{B}\ u(i\Delta) + F\left[y(i\Delta) - H\ \hat{x}(i\Delta) - D\ u(i\Delta)\right] \tag{7-48}$$

The filter gain F can be computed directly from ASYMFIL after ignoring the $\tilde{B}$ and D matrices in the digital plant and observation equations. The matrices $\tilde{A}$ and $\tilde{B}$ follow from EXPINT. The matrix $\tilde{Q}$ can be found from SAMPL using A' in place of A. Finally, we take Q and R to be 4×4 and 3×3 identity matrices, respectively.

The executive program and output data follow.

B/ Executive Program

```
1          PROGRAM KBFIL(INPUT,OUTPUT,TAPE5=INPUT,TAPE6=OUTPUT)
           *******************************************************************
      *
      *
      *    EXECUTIVE PROGRAM FOR KALMAN-BUCY FILTER PROBLEM
      *
      *
           *******************************************************************
           DIMENSION A(16),B(8),H(12),Q(16),R(9),ATIL(16),BTIL(8),QTIL(16),F(
15        112),P(16),DUMMY(114)
           DIMENSION NA(2),NB(2),NH(2),NQ(2),NR(2),NATIL(2),NBTIL(2),NQTIL(2)
          1,IOP(5),NF(2),NP(2),NDUM(2)
           LOGICAL  IDENT,DISC,NEWT,STABLE,FNULL
      C
      C    INPUT HOLLERITH DATA FOR TITLE OF OUTPUT
20         CALL RDTITL
      C
      C    INPUT COEFFICIENT MATRICES AND NOISE INTENSITIES
      C    FOR CONTINUOUS SYSTEM
25         CALL READ(5,A,NA,B,NB,H,NH,Q,NQ,R,NR)
      C
      C    GENERATE COEFFICIENT MATRICES FOR DIGITAL SYSTEM
           IOP(1)=0
           DELT= .05
30         N1=(NA(1)**2) + 1
           CALL EXPINT(A,NA,ATIL,NATIL,DUMMY,NA,DELT,IOP,DUMMY(N1))
           CALL MULT(DUMMY,NA,B,NB,BTIL,NBTIL)
           CALL PRNT(ATIL,NATIL,4HATIL,1)
           CALL PRNT(BTIL,NBTIL,4HBTIL,1)
      C
35    C    COMPUTE VARIANCE MATRIX FOR PROCESS NOISE OF DIGITAL SYSTEM
           CALL EQUATE(Q,NQ,QTIL,NQTIL)
           CALL TRANP(A,NA,DUMMY,NDUM)
           IOP(2)= 0
           CALL SAMPL(DUMMY,NDUM,B,NB,QTIL,NQTIL,R,NR,H,NH,DELT,IOP,DUMMY(N1)
40        1)
           CALL PRNT(QTIL,NQTIL,4HQTIL,1)
```

227

```
            C
            C       SOLVE FOR DIGITAL FILTER GAIN
   45               IDENT = .TRUE.
                    DISC  = .TRUE.
                    STABLE = .FALSE.
                    FNULL  = .TRUE.
                    NEWT   = .TRUE.
   50               ALPHA=.5
                    IOP(1)=1
                    IOP(2)=0
                    IOP(3)=0
                    IOP(4)=0
   55               IOP(5)=0
                    CALL ASYMFIL(ATIL,NATIL,G,NG,H,NH,QTIL,NQTIL,R,NR,F,NF,P,NP,IDENT,
                   1DISC,NEWT,STABLE,FNULL,ALPHA,IOP,DUMMY)
            C
            C
   60               STOP
                    END
```

228

ORACLS   PROGRAM

OUTPUT FOR KALMAN-BUCY FILTER PROBLEM

```
A    MATRIX    4 ROWS         4 COLUMNS
-2.6000000E+00    2.5000000E-01    -3.8000000E+01    0.
-7.5000000E-02   -2.7000000E-01     4.4000000E+00    0.
 7.8000000E-02   -9.9000000E-01    -2.3000000E-01    5.2000000E-02
 1.0000000E+00    7.8000000E-02     0.               0.

B    MATRIX    4 ROWS         2 COLUMNS
 1.7000000E+01    7.0000000E+00
 8.2000000E-01   -3.2000000E+00
 0.               4.6000000E-02
 0.               0.

H    MATRIX    3 ROWS         4 COLUMNS
 1.0000000E+00    0.               0.               0.
 0.               1.0000000E+00    0.               0.
-8.5200000E-02   -4.2100000E-02   -2.2400000E+00   -2.1000000E-03

Q    MATRIX    4 ROWS         4 COLUMNS
 1.0000000E+00    0.               0.               0.
 0.               0.               0.               0.
 0.               0.               1.0000000E+00    0.
 0.               0.               0.               1.0000000E+00

R    MATRIX    3 ROWS         3 COLUMNS
 1.0000000E+00    0.               0.
 0.               1.0000000E+00    0.
 0.               0.               1.0000000E+00

ATIL    MATRIX    4 ROWS         4 COLUMNS
 8.7460216E-01    5.6206553E-02   -1.7644035E+00   -2.3524288E-03
-3.0698872E-03    9.8116490E-01    2.1998505E-01    2.8616468E-04
-3.7743137E-03   -4.8707923E-02    9.7958437E-01    2.5772852E-03
 4.6820759E-02    4.9177112E-03   -4.4809768E-02    9.9996068E-01

BTIL    MATRIX    4 ROWS         2 COLUMNS
 7.9692392E-01    3.2234606E-01
 3.9250218E-02   -1.5895733E-01
 6.1800654E-04    6.8676137E-03
 2.0435169E-02    7.9852845E-03

QTIL    MATRIX    4 ROWS         4 COLUMNS
```

229

OUTPUT FOR KALMAN-BUCY FILTER PROBLEM

9.7864533E-02	-5.6349895E-03	-4.4643769E-02	2.0737669E-03
-5.6349895E-03	4.9959106E-02	4.2255496E-03	-1.2169588E-05
-4.4643769E-02	4.2255496E-03	4.9172184E-02	-6.8173577E-04
2.0737669E-03	-1.2169588E-05	-6.8173577E-04	5.0057551E-02

PROGRAM TO SOLVE THE DISCRETE INFINITE-DURATION OPTIMAL FILTER PROBLEM

A MATRIX   4 ROWS   4 COLUMNS

8.7460216E-01	5.6206553E-02	-1.7644035E+00	-2.3524288E-03
-3.0698872E-03	9.8114690E-01	2.1998505E-01	2.8616468E-04
3.7743137E-03	-4.8707923E-02	9.7958437E-01	2.5772852E-03
4.6820759E-02	4.9177112E-03	-4.4809768E-02	9.9996068E-01

G IS AN IDENTITY MATRIX

H MATRIX   3 ROWS   4 COLUMNS

1.0000000E+00	0.	0.	0.
0.	1.0000000E+00	0.	0.
-8.5200000E-02	-4.2100000E-02	-2.2400000E+00	-2.1000000E-03

INTENSITY MATRIX FOR COVARIANCE OF MEASUREMENT NOISE

R MATRIX   3 ROWS   3 COLUMNS

1.0000000E+00	0.	0.
0.	1.0000000E+00	0.
0.	0.	1.0000000E+00

INTENSITY MATRIX FOR COVARIANCE OF PROCESS NOISE

Q MATRIX   4 ROWS   4 COLUMNS

9.7864533E-02	-5.6349895E-03	-4.4643769E-02	2.0737669E-03
-5.6349895E-03	4.9959106E-02	4.2255496E-03	-1.2169588E-05
-4.4643769E-02	4.2255496E-03	4.9172184E-02	-6.8173577E-04
2.0737669E-03	-1.2169588E-05	-6.8173577E-04	5.0057551E-02

FILTER GAIN

OUTPUT FOR KALMAN-BUCY FILTER PROBLEM

F   MATRIX   4 ROWS   3 COLUMNS
```
 5.3454233E-01  -1.0000879E-02   3.6312717E-01
-4.3653081E-02   1.8601819E-01  -3.4324468E-02
-8.4771306E-02  -9.0256464E-03  -1.2135706E-01
-7.0927842E-03   3.8165991E-02  -1.4113759E-01
```

STEADY-STATE VARIANCE MATRIX OF RECONSTRUCTION ERROR

P   MATRIX   4 ROWS   4 COLUMNS
```
 8.8649680E-01  -5.7194386E-02  -2.2094025E-01  -1.2626090E-01
-5.7194386E-02   2.3575573E-01   8.3030258E-03   5.2808810E-02
-2.2094025E-01   8.3030258E-03   1.0585196E-01   9.1725042E-02
-1.2626090E-01   5.2808810E-02   9.1725042E-02   1.9945248E+01
```

EIGENVALUES OF P

EVLP MATRIX   4 ROWS   1 COLUMNS
```
4.7257495E-02
2.3111030E-01
9.4813881E-01
1.9946666E+01
```

A-FH MATRIX   4 ROWS   4 COLUMNS
```
3.7098027E-01   8.1495086E-02  -9.5098664E-01  -1.5898617E-03
3.7658749E-02   7.9368365E-01   1.4309824E-01   2.1408330E-04
7.8205998E-02  -4.4791409E-02   7.0774455E-01   2.3224354E-03
4.1888621E-02  -3.9190172E-02  -3.6095797E-01   9.9966429E-01
```

EIGENVALUES OF A-FH MATRIX

```
5.4024239E-01   2.2217840E-01
5.4024239E-01  -2.2217840E-01
7.9425006E-01   0.
9.9735592E-01   0.
```

231

VI/ EIGENVALUE PLACEMENT

A/ Problem Statement

This problem illustrates the eigenvalue placement procedure. The plant equations are obtained from a linearized mathematical model of the longitudinal dynamics of an F-8 aircraft [7-1]. The system can be written as (7-2) with

$$
A = \begin{bmatrix}
-0.49 & 0.00005 & -4.8 & 0.0 \\
0.0 & -0.015 & -14.0 & -32.2 \\
1.0 & -0.00019 & -0.84 & 0.0 \\
1.0 & 0.0 & 0.0 & 0.0
\end{bmatrix}
\tag{7-49}
$$

and

$$
B = \begin{bmatrix}
-8.7 \\
-1.1 \\
-0.11 \\
0.0
\end{bmatrix}
\tag{7-50}
$$

The open-loop eigenvalues of A are

$$
(-0.0068517471 \pm j\ 0.076518821)
\tag{7-51}
$$

and

$$
(-0.66564825 \pm j\ 2.1821221)
\tag{7-52}
$$

It is required to use state variable feedback to modify A so that the
real parts of (7-51) and (7-52) are reduced by 20 percent while the
imaginary parts are unchanged. In other words, find F in the control
law

$$u = -Fx \qquad\qquad (7\text{-}53)$$

such that the closed-loop response matrix A-BF has eigenvalues

$$(-0.00822209 \pm j\ 0.076518821) \qquad\qquad (7\text{-}54)$$

and

$$(-0.7987779 \pm j\ 2.1821221) \qquad\qquad (7\text{-}55)$$

The (A,B) pair can be shown to be completely controllable whereby
the F matrix can be calculated directly from subroutine POLE of
ORACLS. The executive program and output data follow.

B/ Executive Program

```
 1          PROGRAM EIPLAC(INPUT,OUTPUT,TAPE5=INPUT,TAPE6=OUTPUT)

            ****************************************************************
 5          *
            *
            *     EXECUTIVE PROGRAM FOR EIGENVALUE PLACEMENT PROBLEM
            *
            *
10          ****************************************************************

            DIMENSION A(4,4),B(4),EVAL(4),F(4),DUMMY(108)
            DIMENSION NA(2),NB(2),NF(2)
15    C
      C     INPUT HOLLERITH DATA FOR TITLE OF OUTPUT
      C     CALL RDTITL

      C
20    C     DEFINE A AND B MATRICES
      C
            NA(1) = 4
            NA(2) = 4
            NB(1) = 4
25          NB(2) = 1
            CALL NULL(A,NA)
            A(1,1) = -.49
            A(3,1) = 1.0
            A(4,1) = 1.0
30          A(1,2) = .00005
            A(2,2) = -.015
            A(3,2) = -.00019
            A(1,3) = -4.8
            A(2,3) = -14.0
35          A(3,3) = -.84
            A(2,4) = -32.2
            B(1) = -8.7
            B(2) = -1.1
            B(3) = -.11
40          B(4) = 0.0
      C
      C
```

234

```
      C         SPECIFY PRESCRIBED EIGENVALUES
 45             EVAL(1) = -.00822209
                EVAL(2) = .076518821
                EVAL(3) = -.7987779
                EVAL(4) = 2.1821221
                NUMR = 0

      C
      C         SPECIFY PRINT OPTION
 50             IOP = 1

      C
      C         FIND GAIN TO PLACE EIGENVALUES
 55             CALL POLE(A,NA,B,NB,EVAL,NUMR,F,NF,IOP,DUMMY)

      C
      C
 60             STOP
                END
```

OUTPUT FOR EIGENVALUE PLACEMENT PROBLEM

GIVEN THE COMPLETELY CONTROLLABLE PAIR (A,B), THE MATRIX F IS COMPUTED
SUCH THAT A-BF HAS PRESCRIBED EIGENVALUES

```
    A    MATRIX      4 ROWS        4 COLUMNS
-4.9000000E-01   5.0003000E-05  -4.8000000E+00   0.
 0.             -1.5000000E-02  -1.4000000E+01  -3.2000000E+01
 1.0000000E+00  -1.9000000E-04  -8.4000000E-01   0.
 1.0000000E+00   0.              0.              0.

    B    MATRIX      4 ROWS        1 COLUMNS
-8.7000000E+00
-1.1000000E+00
-1.1000000E-01
 0.
```

THE PRESCRIBED EIGENVALUES ARE STORED IN EVAL
PAST ENTRY   O NUMBERS CORRESPOND TO REAL AND IMAGINARY PARTS

```
   EVAL  MATRIX      4 ROWS        1 COLUMNS
-8.2220900E-03
 7.6518821E-02
-7.9877790E-01
 2.1821221E+00
```

DENOTE CHARACTERISTIC POLYNOMIAL OF A BY D(S)
       N   N-1   N-2
D(S)=S +S   D(1)+S   D(2)+...+SD(N-1)+D(N)

AL = COL.(D(N),D(N-1),...,D(1))

    AL   MATRIX   4 ROWS   1 COLUMNS

OUTPUT FOR EIGENVALUE PLACEMENT PROBLEM

```
3.0718800E-02
7.9180600E-02
5.2288900E+00
1.3450000E+00
```

FOR THE   4 EIGENVALUES STORED IN EVAL THE CHARACTERISTIC POLYNOMIAL IS

$$D(S) = S^{N} + S^{N-1} D(1) + S^{N-2} D(2) + \ldots + SD(N-1) + D(N)$$

ALTL = COL.(D(N),D(N-1),...,D(1))

ALTL MATRIX   4 ROWS   1 COLUMNS
```
3.1980998E-02
9.8255584E-02
5.4318962E+00
1.6140000E+00
```

F   MATRIX   1 ROWS   4 COLUMNS
```
-3.0962910E-02   5.8292950E-06   3.3720374E-03   -2.1331284E-03
```

A-BF MATRIX   4 ROWS   4 COLUMNS
```
-7.5937732E-01   1.0071487E-04   -4.7706633E+00   -1.8558217E-02
-3.4059201E-02   -1.4993588E-02   -1.3996291E+01   -3.2202346E+01
 9.9659408E-01   -1.8935878E-04   -8.3962908E-01   -2.3464412E-04
 1.0000000E+00   0.              0.              0.
```

EIGENVALUES(VAL MATRIX) AND EIGENVECTORS(VECT MATRIX) OF A-BF

VAL MATRIX   4 ROWS   2 COLUMNS
```
-8.2220900E-03   7.6518821E-02
-8.2220900E-03   -7.6518821E-02
-7.9877790E-01   2.1821221E+00
-7.9877790E-01   -2.1821221E+00
```

VECT MATRIX   4 ROWS   4 COLUMNS
```
 1.8164153E-04   2.7795595E-05   1.1523450E-01   0.
 9.9132426E-01   1.3141722E-01   8.2983089E-01   -5.4118489E-01
-7.6420695E-04   4.7634686E-06   1.0355453E-03   -5.2539029E-02
 1.0694608E-04   -2.3853067E-03   -1.7046635E-02   -4.6568439E-02
```

237

VII/ TRANSFER MATRIX

A/ Problem Statement

This problem illustrates the calculation of the elements of the transfer

matrix for the system

$$\dot{x}(t) = A\ x(t) + B\ u(t)$$
$$y(t) = H\ x(t) \tag{7-56}$$

with

$$A = \begin{bmatrix} -2.60 & 0.25 & -38.0 & 0.0 \\ -0.075 & -0.27 & 4.40 & 0.0 \\ 0.078 & -0.99 & -0.23 & 0.052 \\ 1.0 & 0.078 & 0.00 & 0.00 \end{bmatrix} \tag{7-3}$$

$$B = \begin{bmatrix} 17. & 7.0 \\ 0.82 & -3.2 \\ 0.0 & 0.046 \\ 0.0 & 0.0 \end{bmatrix} \tag{7-4}$$

and

$$H = \begin{bmatrix} 1.0 & 0.0 & 0.0 & 0.0 \\ 0.0 & 1.0 & 0.0 & 0.0 \\ -0.0852 & -0.0421 & -2.24 & -0.0021 \end{bmatrix} \tag{7-46}$$

The subroutine LEVIER of ORACLS can be directly applied. The executive

program and output data follow.

Based on the output data

$$H(sI-A)^{-1}B = \frac{\begin{bmatrix} 17s^3+8.70s^2+106.s-.430, & 7s^3+.952s^2-90.06s+.368 \\ .82s^3+1.046s^2+8.46s+5.51, & -3.2s^3-9.38s^2-8.46s-4.72 \\ -1.48s^3-2.056s^2-10.4s-1.026, & -.572s^3-8.93s^2-14.11s+.308 \end{bmatrix}}{(s^4 + 3.1s^3 + 8.7s^2 + 16.99s + .418)} \cdot$$

$$\tag{7-57}$$

B/ Executive Program

```
  1            PROGRAM TRANFR(INPUT,OUTPUT,TAPE5=INPUT,TAPE6 =OUTPUT)
       ***************************************************************
       *
  5    *       EXECUTIVE PROGRAM FOR TRANSFER MATRIX PROBLEM
       *
       ***************************************************************

 10            DIMENSION A(4,4),B(4,2),H(3,4),C(80),HCB(24),D(4),DUMMY(32)
               DIMENSION NA(2),NB(2),NH(2),NC(2),NHCB(2),ND(2)
 15            LOGICAL BIDENT,HIDENT

       C
       C       INPUT HOLLERITH DATA FOR TITLE OF OUTPUT
       C       CALL ROTITL
 20
       C
       C       DEFINE A,B,H MATRICES
       C
               NA(1) = 4
 25            NA(2) = 4
               NB(1) = 4
               NB(2) = 2
               NH(1) = 3
               NH(2) = 4
               CALL NULL(A,NA)
 30            A(1,1) = -2.60
               A(1,2) =  0.25
               A(1,3) = -38.0
               A(2,1) = -0.075
               A(2,2) = -0.27
               A(2,3) =  4.4
 35            A(3,1) =  0.078
               A(3,2) = -0.99
               A(3,3) = -0.23
               A(3,4) =  0.052
               A(4,1) =  1.0
 40            A(4,2) =  0.078
               CALL NULL(B,NB)
```

```
45          B(1,1) = 17.0
            B(1,2) = 7.0
            B(2,1) = 0.82
            B(2,2) = -3.2
            B(3,2) = 0.046
            CALL NULL(H,NH)
            H(1,1) = 1.0
50          H(2,2) = 1.0
            H(3,1) = -0.0852
            H(3,2) = -0.0421
            H(3,3) = -2.4
            H(3,4) = -0.0021

      C
55    C     DEFINE LOGICAL INPUT
            BIDENT = .FALSE.
            HIDENT = .FALSE.
      C
      C
60    C     SPECIFY PRINT OPTION
            IOP = 1
      C
      C
65    C     CALCULATE ELEMENTS OF TRANSFER MATRIX
            CALL LEVIER(A,NA,B,NB,H,NH,C,NC,HCB,NHCB,D,ND,BIDENT,HIDENT,IOP,
           1DUMMY)
      C
70    C     STOP
            END
```

240

C/ Output from ORACLS

OUTPUT FOR TRANSFER MATRIX PROBLEM

THE TRANSFER MATRIX H((SI-A)INVERSE)B IS COMPUTED
FOR THE (A,B,H) SYSTEM

A   MATRIX   4 ROWS   4 COLUMNS
-2.6000000E+00   2.5000000E-01 -3.8000000E+01   0.
-7.5000000E-02 -2.7000000E-01   4.4000000E+00   5.2000000E-02
 7.8000000E-02 -9.9000000E-01 -2.3000000E-01   0.
 1.0000000E+00   7.8000000E-02   0.   0.

B   MATRIX   4 ROWS   2 COLUMNS
1.7000000E+01   7.0000000E+00
6.2000000E-01 -3.2000000E+00
0.   4.6000000E-02
0.   0.

H   MATRIX   3 ROWS   4 COLUMNS
1.0000000E+00   0.   0.   0.
0.   1.0000000E+00   0.   0.
-8.5200000E-02 -4.2100000E-02 -2.4000000E+00 -2.1000000E-03

$$(SI-A)INVERSE=\frac{(S^{N-1} C(0)+S^{N-2} C(1)+...+SC(N-2)+C(N-1))}{D(S)}$$

$$D(S)=S^N+S^{N-1} D(1)+S^{N-2} D(2)+...+SD(N-1)+D(N)$$

MATRIX C(0) IN (SI-A) INVERSE EQUATION

1.0000000E+00   0.   0.   0.
0.   1.0000000E+00   0.   0.
0.   0.   1.0000000E+00   0.
0.   0.   0.   1.0000000E+00

OUTPUT FOR TRANSFER MATRIX PROBLEM

MATRIX HC(0)B

```
1.7000000E+01    7.0000000E+00
8.2000000E-01   -3.2000000E+00
-1.4829220E+00  -5.7208000E-01
```

MATRIX C(1) IN (SI-A) INVERSE EQUATION

```
5.0000000E-01   2.5000000E-01   -3.8000000E+01   0.
-7.5000000E-02  2.8300000E+00    4.4000000E+00   0.
7.8000000E-02  -9.9000000E-01    2.8700000E+00   5.2000000E-02
1.0000000E-02   7.8000000E-02    0.              3.1000000E+00
```

MATRIX HC(1)B

```
8.7050000E+00    9.5200000E-01
1.0456000E+00   -9.3786000E+00
-2.0556001E+00  -8.9308952E+00
```

MATRIX C(2) IN (SI-A) INVERSE EQUATION

```
4.4181000E+00   3.7677500E+01   -9.1600000E+00   -1.9760000E+00
3.2995000E-01   3.5620000E+03    1.4290000E+01    2.2880000E-01
1.4731000E-01  -2.6504440E+00    7.2075000E-01    1.4924000E-01
4.9415000E-01   4.7074000E-01   -3.7658800E+01    8.7008500E+00
```

MATRIX HC(2)B

```
1.0600325E+02   -9.0062660E+01
8.4619900E+00   -8.4594100E+00
-1.0397153E+01  -1.4112772E+01
```

OUTPUT FOR TRANSFER MATRIX PROBLEM

MATRIX C(3) IN (SI-A) INVERSE EQUATION

```
-1.7846400E-02   -1.5412800E-01    2.1174174E-12   -4.7632000E-01
 2.2880000E-01    1.9760000E+00   -4.1211479E-13    7.4308000E-01
 1.3735800E-02    2.3545600E-02   -1.1368684E-13    3.7479000E-02
 4.4435241E+00    3.7955336E+01   -8.0453800E+00    1.5027353E+01
```

MATRIX HC(3)B

```
-4.2977376E-01    3.6828480E-01
 5.5099200E+00   -4.7216000E+00
-1.0261022E+00    3.0798750E-01
```

ERROR IN SATISFYING CAYLEY-HAMILTON THEOREM

```
     EROR  MATRIX          4 ROWS          4 COLUMNS
 5.2047255E-13   -2.0261570E-12    7.9258200E-12    1.1010570E-13
-3.5527137E-15    2.6645353E-14   -6.4417804E-13   -2.1429969E-14
-4.4408921E-15    1.0214052E-13   -1.3145041E-13   -5.9117156E-15
-3.4106051E-13   -2.5579538E-13    1.3429258E-12   -4.2099657E-13
```

```
     D  MATRIX          4 ROWS          1 COLUMNS
 3.1000000E+00
 8.7008500E+00
 1.6985506E+01
 4.1835976E-01
```

REFERENCES

7-1. J. R. Elliott, "NASA's Advanced Control Law Program for the F-8 Digital Fly-by-Wire Aircraft," IEEE Trans. Autom. Control, vol. AC-22, no. 5, Oct. 1977, pp. 753-756.

7-2. G. Stein and A. H. Henke, A Design Procedure and Handling-Quality Criteria for Lateral-Directional Flight Control Systems, AFFDL-TR-70-152, U.S. Air Force, May 1971.

7-3. H. Kwakernaak and R. Sivan, Linear Optimal Control Systems, John Wiley and Sons, Inc., c.1972.